AI, Human, and Non-Human Consciousness

Our Digital Conscious Metaverse

S.R. Joye, BSEE, MA, PhD

Published by the Viola Institute

Assisi, Italy
2026

This book is dedicated to my friend and mentor, the pioneer of consciousness studies, Leslie Allan Combs, and to John Lilly, my earliest mentor in the field of electrical engineering and the isolation tank exploration of consciousness,
and to my son, Jason, and to my daughter, Alyssa.

Some of the material in this book has appeared in the author's previous books: *Developing Supersensible Perception, The Electromagnetic Brain, Tuning the Mind, The Little Book of Consciousness,* and *The Little Book of the Holy Trinity.*

Printed in the United States of America

AI, Human, and Non-Human Consciousness:
Our Digital Conscious Metaverse

ISBN-13: 978-1-950761-34-0

OTHER BOOKS BY THE SAME AUTHOR

The Electromagnetic Brain: EM Field Theories on the Nature of Consciousness

Tuning the Mind: Geometries of Consciousness

Exploring the Noosphere: Teilhard de Chardin

The Little Book of Consciousness: Holonomic Brain Theory and the Implicate Order

Sub-Quantum Consciousness: A Geometry of Consciousness Based Upon the Work of Karl Pribram, David Bohm, and Pierre Teilhard de Chardin

Contents

Our Digital Conscious Metaverse

KEY TERMS AND DEFINITIONS

Explicate Order: One of the two primary dimensional orders in the universe, as posited by David Bohm. The explicate order encompasses spacetime and all manifestations within space- time. The other dimensional order Bohm termed the "Implicate Order."

Holosphere: A spherical boundary that has the diameter of one Planck length (3 x 10^{-35} meter). It is the event horizon between the explicate order (spacetime) and the implicate order (hidden dimensions).

Holoplenum: A continuous plenum of Planck length holospheres at the bottom of space.

Holoflux: Multidimensional holographic energy of information-consciousness. Holoflux in spacetime is projecting out of dimensions within Bohm's implicate order; holoflux is synonymous with the electromagnetic field in spacetime, resonating with multiple fields throughout other dimensions of the implicate order.

Implicate Order: One of two primary dimensional orders in the universe, as posited by David Bohm. The implicate order is a nonlocal, nontemporal domain that can be found below the Planck length. In holoflux theory it is equivalent to the frequency-phase domain in communications engineering. Bohm felt that the implicate is "somehow 'deeper' and more fundamental than the explicate" (Peat, 1987, p. 262).

Metaverse: The entirety of the conscious universe; an integrated network of multiple dimensions that make up the whole of reality. This includes, in addition to space and time, a minimum of at least an additional seven hidden dimensions (as predicted by string theory) as well as innumerable virtual reality dimensions open to creation, development, and exploration by consciousness.

Planck length: The smallest possible length in space, determined by Max Planck from the gravitational constant and the speed of light. The value of the Planck length is 3×10^{-35} meter.

Planck time: The smallest possible unit of time, determined by Max Planck from the gravitational constant and the speed of light. The value of Planck time is 5×10^{-44} sec.

Psychonauts: The smallest possible unit of time, determined by Max Planck from the gravitational

Introduction

This book proposes a model of brain–mind function grounded in quantum holographic and holonomic principles, extending established neuroscientific theories of distributed information processing. Building on Pribram's holonomic brain theory and Bohm's holographic quantum framework, we introduce the concept of a *holoflux*: a dynamic, generative informational process through which neural activity patterns and conscious experience emerge. In this view, brain function is not reducible to localized neural mechanisms alone, but arises from distributed interference-like patterns that encode information nonlocally while remaining operationally expressed through classical neural networks.

Within this framework, quantum holographic dynamics are conceived as an active informational substrate influencing large-scale neural coherence, integration, and pattern formation. This enables a dual mode of organization: locally, through Newtonian-mechanistic neural interactions, and globally, through nonlocal holographic informational correlations. Such a *holo-informational* architecture provides a theoretical basis for understanding integrative cognitive phenomena—including perceptual unity, memory retrieval, and conscious awareness—as emergent properties of distributed neural–informational processes rather than solely synaptic localization.

By invoking the mathematical properties of holographic systems, in which information about the whole is distributed across each component, this model suggests that individual brain–mind systems participate in a larger informational continuum. While remaining compatible with current neuroscientific observations, the framework opens the possibility that neural holographic dynamics are embedded within a universal informational field, allowing brain function and consciousness to be understood as expressions of an irreducible, globally distributed informational order.

The Granularity of Space and Time: Our Digital Cosmos

When I was a young student, I was intrigued by a lecture I heard by the California physics professor, Richard Feynman. He asked us to consider the following. If we were to take a hammer and try to strike an anvil with it, we could mathematically track the distance of the hammer from the anvil, at each moment of time. At some moment the hammer would reach half of the remaining distance to the anvil. You could plot the time of the next moment after that, when the hammer would again reach half of the distance remaining. If space were continuous, you could continue measuring the time at which the hammer managed to move closer to the anvil by halving the distance. But, if space were continuous, you would you never actually be able to strike the anvil because you could continue to move half the previous distance closer to the anvil. Thus, space is not continuous, there must be a final limit, so space must be granular, not continuous. This goes for time too, he said. Time is not continuous, but is granular. Then, having duly confused us students, Feynman would launch into his lecture on quantum physics.

Quantum means "discrete" in one sense, i.e. there is no continuity but a sort of "jump" (in fact the phrase "quantum jump" is often used). This "jump" is due to the bottom limits to space and time that were discovered to be the solution to solve a major problem that had plagued physicists for decades, i.e. the mathematics to model the relationship of temperature peaks to the peak wavelength of the emf in the infrared band (that we sense and term "heat").

Max Planck (known as "the father of Action") somehow intuited the solution that had escaped everyone else for decades. He had been working to solve the puzzle of the emission of radiation from heated objects, often called the problem of "black-body radiation."

All normal matter at temperatures above absolute zero emits electromagnetic radiation. It had long been noticed that charcoal could be seen to be giving off different colors that depended directly upon the temperature to which they had been heated. As the temperature of charcoal is increased, the visible emission began to be seen as a dull red, then as the temperature increased, it became more orange, then yellow, then green, and finally violet. Laboratory measurements had been made of the wavelength of light coming from glowing charcoal at different temperatures (fig. 1), but the calculations made (using "Classical theory," prior to quantum theory), shown in the dashed curve) were way off, and never matched the actual 5000 K measurements, shown in the solid line curves. In the figure can be seen the large discrepancy for a temperature of 5000 Kelvin.

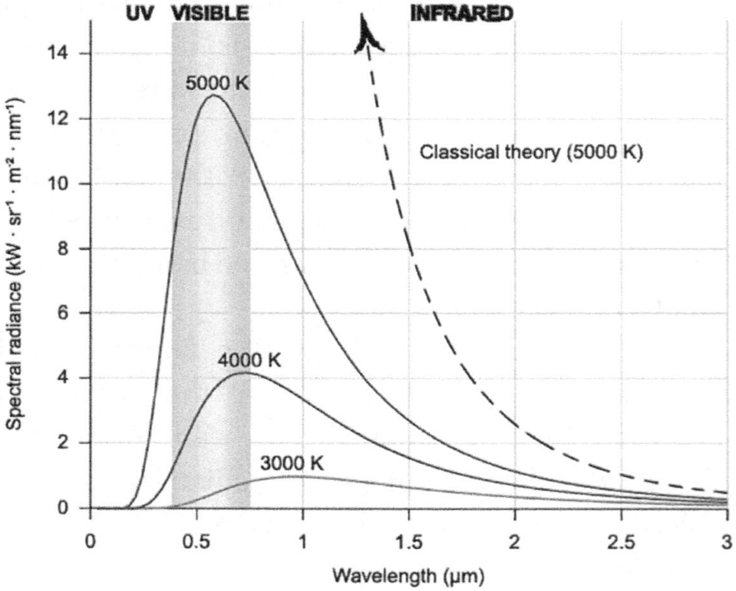

Fig. 1. Black-Body Radiation Curve Discrepancy[1]

Until Planck solved the puzzle, nobody had been able to accurately calculate the "black-body radiation curves" that would match laboratory measurements.

Planck's intuitive flash of insight was this:

1.) There can be no completely continuous Actions in space and time.
2.) There is both a top and bottom limit to the dimensions of time and space.
3.) Action must thus be "granular" (corollary: the universe is digital)

Knowing that the speed of light is invariant, Planck immediately sought to calculate the bottom limit of space, spending months in searching among various mathematical configurations to use the known constants of the speed of light and the gravitational constant in order to match exactly the black-body radiation curves. Eventually they all came together in his famous equation, which defines what is now called the Planck Length (fig. 2).

$$\ell_P = \sqrt{\frac{\hbar G}{c^3}}$$

ℓ_P = **The Planck Length**
$\hbar$ = ***Planck's constant***
G = ***Gravitational constant***
c = ***Speed of light***

Fig. 2. The Planck Length Equation

Because length divided by velocity results in time, he was also able to calculate the fundamental unit of time (fig. 3) as well as several other fundamental lower limits.

Calculation of the Planck length

$$\ell_P = \sqrt{\frac{\hbar G}{c^3}} = 1.616229 \times 10^{-35} \text{ meter}$$

Calculation of the Planck time

$$t_P = \frac{\ell_P}{c} = \sqrt{\frac{\hbar G}{c^5}} = 5.391247 \times 10^{-44} \text{ seconds}$$

Fig. 3. Calculation of the Planck length and the Planck time constants

These then are the lower limits to our space and time dimensions:

Planck length constant	➔ 1.616255×10^{-35} Meters
Planck time constant	➔ 5.391247×10^{-44} Seconds
Planck mass constant	➔ 2.176434×10^{-8} Kg
Planck temperature	➔ 1.416784×10^{32} Kelvin

Note that at the smallest dimension, the Planck length, the temperature reaches the maximum possible in spacetime , the Planck temperature constant. That is because at smaller dimensions, the radii of the spinning energy "strings" is smaller, and thus the frequencies of the emitted radiation are higher. Higher frequencies emit radiation at higher energy levels.

Henceforth, using these constants in the mathematical model made the curve calculations match the observed curve exactly. A year after his discovery, Max Plank began using the term "quanta" (Latin: "how much" or "amount") to describe these values, and some years later, his colleagues began to call these these the "Planck constants." In 1918 Planck was awarded the Nobel Prize for his discovery of these constants.

The Planck time constant of 5.391247×10^{-44} seconds is the smallest unit of time possible in our spacetime continuum. While humans are familiar with the duration of "one human second" of time, conventionally measured with our own time keeping clock devices, it is difficult to imagine much smaller time units. One of the fastest contemporary Intel microprocessor chips runs at 4.7 GHz, which is 4.7×10^{9} . However within this "one human second" there are 10^{9} "processor ticks" and 10^{44} "Planck time ticks" per second, which is truly mind-boggling!

	Scientific Notation	Seconds, in Standard Notation
Standard "second"	10^{0}	1 (note that this sets the scale of human-perceived time to approximately equal the average human heartbeat)
Early computer processor (Z80 chip)	10^{-5}	0.00001
Intel microprocessor (Intel Core i5 chip)	10^{-9}	0.000000001
Planck time	10^{-44}	0.000000000000000000000 0000000000000000000001

Comparison of time units:
Scientific notation and Standard notation.

Because time does not exist below this lowest granular division of time, Planck time, there is a quantum "jump" as time flows, a jump from one tick to the next. But can we imagine what occurs between the two ticks? In between these ticks, time does not exist. This is what metaphysicians would call the void, the timeless eternal, the transcendent. In the cosmological physics of David Bohm, it is the implicate order, the ground of being *into which* all information folds inward from our external spacetime , and from *out of which* all things are unfolded and projected into spacetime by the same implicate order.

To Bohm, all particles that exist within our cosmos of spacetime must be considered to be actually projections out of a higher dimensional order:

> We may thus regard each of the 'particles' constituting a system as a projection of a 'higher-dimensional' reality, rather than as a separate particle, existing together with all the others in a common three-dimensional space. For example, in the experiment of Einstein, Podolsky and Rosen, each of two atoms that initially combine to form a single molecule are to be fully regarded as three-dimensional projections of a six-dimensional reality.[2]

Pribram and Bohm spoke about a holomovement occurring outside of the four dimensions of space and time. Pribram did not like the use of the word "movement" in this naming convention, and recommended to Bohm that they use the term "holoflux," which is not a movement in time, but more of an evolution of information within the frequency domains of the implicate order. In fact one of the attributes of the implicate order, as Bohm understood it, would be a continuous *enfolding* of information from the spacetime domain, back into the hidden dimensions of the implicate order, where somehow the total information coming from spacetime for that one "tick" of time would be immediately superimposed over all of the pre-existing information stored in the implicate order (Akasha, according to Bucke, Steiner, and Laszlo). At the very next quantum "tick" of time, the cosmos would be reprojected into spacetime from information holoflux within the implicate order. The cycle can truly be seen to be a digital process, with the cosmos being created and recreated with each tick of the Planck-time clock. Bohm says:

> What follows from all this is that basically the implicate order has to be considered as a process of enfoldment and unfoldment in a higher-dimensional space. Only under certain conditions can this be simplified as a process of enfoldment and unfoldment in three dimensions.[3]

Bohm goes on to describe this projected information holoflux as it appears within the constricted spacetime dimensions:

> The electromagnetic field, which is the ground of the holographic image, obeys the laws of the quantum theory, and when these are properly applied to the field it is found that this, too, is actually a multidimensional reality which can only under certain conditions be simplified as a three-dimensional reality. Quite generally, then, the implicate order has to be extended into a multidimensional reality. In principle this reality is one unbroken whole, including the entire universe with all it's 'fields' and 'particles.'[4]

The Pribram-Bohm Holoflux Topology

The psychophysical model of consciousness which is developed here is presented in fig. 4, the "Pribram-Bohm Holoflux Model," where the basic theory is diagrammed as consciousness transforming between nonlocal and local regions of experience and information.

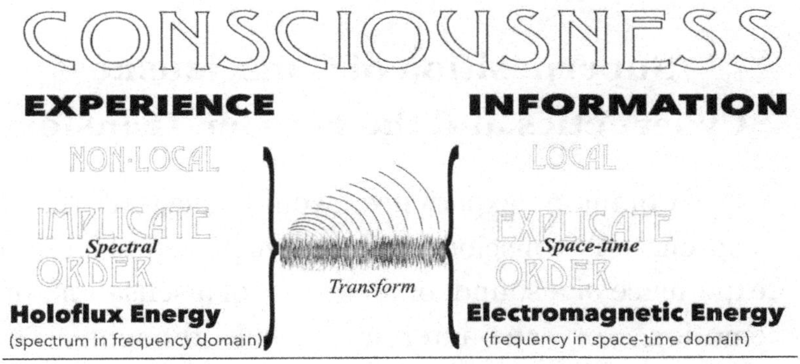

Fig. 4. The Pribram-Bohm Holoflux Model

The Pribram-Bohm hypothesis regards consciousness as a cybernetic energy process, a holoflux transforming between two orders of being in "an undivided flowing movement without borders" (Bohm, 1980, p. 172). To the left in the diagram of Figure 1, consciousness is expressed as a spectrum of holoflux energy in Bohm's implicate order. This holoflux energy resonates with electromagnetic energy of the same frequencies to the right in the diagram, in the spacetime region, or explicate order (Bohm, 1980, p. 159).

Viewed from left to right, the diagram reveals a spectrum of holoflux energy in the transcendental implicate order transforming and translated into "things" and "events" in local spacetime , and conversely, viewing the diagram from right to left, information generated by "things" and "events" interacting throughout spacetime is seen to be transforming (folding) back into the implicate order. The process is described as a continuous cybernetic cycle, perhaps occurring at a regular clock-rate.

Superposition of Consciousness Cybernetics and the Fourier Transform

Common experience would suggest that all awareness is consciousness *of* something, such as the experience of a sound, of an image, of a sensation, of an emotion, of an interior verbal thought. These experiences seem to be superpositioned, they often occur at what seem to be the same perceptual moment. Yet each simultaneous stream of experience remains distinct, somehow integrated with all the others. But what is it that is "looking at" this stream of experiences? It is as if there is some meta-consciousness that is more than the sum of each of these individual "experience streams," per se, but rather some other, more comprehensive level, some panoramic perspective that is able to embrace and comprehend them all, and which has the amazing ability to *fine tune* its own selected focus upon *one or more* of these streams of awareness while simultaneously dampening and filtering out the many others.

In signal analysis this stream phenomenon is explained by the *superposition principle*, formalized in 1822 by the French mathematician Jean-Baptiste Fourier, who developed the mathematics of what is now called "Fourier analysis" during his search for a mathematical relationship between spacetime and frequency (Feynman, Leighton, & Sands, 1964, p. 286). Because signals are more readily superpositioned and manipulated (filtered, amplified, etc.) within the *frequency domain* than in the *time domain*, the Fourier transform equations have become primary and ubiquitous mathematical tools in physics and engineering for analyzing, synthesizing, and transmitting signals between two domains:

1. a "spacetime domain (t_d)," and

2. a "frequency domain (f_d)."

Much of electrical engineering circuit design is done within the frequency domain, and only subsequently implemented with time domain components, as described here by Francis F. Kuo, chief electrical engineer at the original Bell Telephone Laboratory from which emerged the transistor, the laser, and radio astronomy, information theory, and the Unix operating system. In his textbook on *Network Analysis and Synthesis*, Kuo (1962) states:

> We see that in the **time domain** (i.e., where the independent variable is **t**) the voltage–current relationships are given in terms of differential

> equations. On the other hand, in the complex **frequency domain**, the voltage–current relationships for the elements are expressed in **algebraic** equations. Algebraic equations are, in most cases, more easily solved than differential equations. Herein lies the **raison d'être** for describing signals and networks in the frequency domain as well as in the time domain. (p. 13)

Norbert Wiener (1948) coined the term "cybernetics" from the Greek κυβερνήτης— "steersman, governor, pilot, or rudder" (p. 11)—during his own work at the same Bell Telephone Laboratory as Kuo, and made use of Fourier's transform to model and analyze brain waves in the frequency domain, where he discovered clear evidence of "self-organization of electroencephalograms or brain waves" (p. 181). Using Fourier analysis, an approach which later became of great interest to Bohm, Wiener (1948) was able to detect uniquely narrow frequency ranges, centered within different spatial locations on the cortex, that repeatedly exhibited auto- correlation (p. 191). Regions on the cortex were identified where specific ranges of frequencies were found to coalesce toward intermediate frequencies, seeming both to attract and to strengthen one another, exhibiting *resonance* or "self tuning" to amplify and consolidate signals into narrowly specific ranges in the frequency domain f_d (p. 198). His research led Wiener to conjecture that the *infrared band* of electromagnetic flux may be the loci of "self–organizing systems":

> We thus see that a nonlinear interaction causing the attraction of frequency can generate a self-organizing system, as it does in the case of the brain waves we have discussed. This possibility of self-organization is by no means limited to the very low frequency of these two phenomena. Consider self–organizing systems at the frequency level, say, of infrared light. (p. 202)

Three years after Wiener's (1948) publication of *Cybernetics*, David Bohm (1951) stressed the importance of Fourier's equations on the first page of his well-received 646-page textbook, *Quantum Theory*, where he encouraged a familiarity with Fourier analysis for an ontological understanding of quantum phenomena:

> It seems impossible to develop quantum concepts extensively without Fourier analysis. It is, therefore, presupposed that the reader is moderately familiar with Fourier analysis. (p. 1)

For purposes of this discussion, the basic understanding of Fourier analysis is simply that *frequency vibrations* manifest within two distinct dimensions or domains: a spacetime domain and a frequency domain. Until recently, physicists have focused exclusively within spacetime to conduct their research, considering only space and time as having any "reality" and considering the ontological reality of the frequency domain, if at all, in the same vague category as the domain of mathematics itself (i.e., in some unspecified transcendent dimension). Whether there might somehow exist a "real" dimension *outside of* spacetime , or *beyond* spacetime has generally been beyond the purview of the physical sciences. Yet the experienced reality of a region of consciousness beyond spacetime is supported by the vast body of first–hand reports generated by religious, mystical, or near-death experiences. In an approach to such experiences, William James (1909), the "father of American psychology," writes:

> The further limits of our being plunge, it seems to me, into an altogether other dimension of existence from the sensible and merely "understandable" world. Name it the mystical region, or the supernatural region, whichever you choose. So far as our ideal impulses originate in this region (and most of them do originate in it, for we find them possessing us in a way for which we cannot articulately account), we belong to it in a more intimate sense than that in which we belong to the visible world, for we belong in the most intimate sense wherever our ideals belong. (p. 318)

Fourier's transform equations (fig. 5) between the two domains of time (t_d) and frequency (f_d) are more than simply mathematical equations, written down as functions in the abstract symbolic language of calculus (Stein & Shakarchi, 2003, pp. 134–36).

$$f(t)=\int_{-\infty}^{+\infty} X(F)e^{j2\pi Ft}\,dF \qquad f(F)=\int_{-\infty}^{+\infty} x(t)e^{-j2\pi Ft}\,dt$$

Fourier integral transform of a continuous frequency function into the time domain (t_d).

Fourier integral transform of a continuous time function into the frequency domain (F_d).

Fig. 5. The Fourier transform and inverse transform

These two expressions indicate that any function in the timespace domain can be transformed into and expressed equivalently as an infinite series of frequency spectra functions in the frequency domain. The transformation is also possible in the opposite direction, such that any arbitrary function in the frequency domain, can be transformed into and expressed by an infinite series of time functions. The two domains mirror one another.

Beyond purely mathematical considerations, the equations can be taken as models of an actual cosmic process (i.e., much as Newton's Law model the phenomenon of gravity) and they can be understood as mirroring the cosmos in mathematical terms. The model of consciousness presented in this paper proposes that there is indeed an ontological reality to this other region, and that this region is synonymous with Bohm's "implicate order," Pribram's "holonomic frequency domain."

Pribram's Holonomic Mind/Brain Theory

The neurosurgeon Karl Pribram (1971) was one of the first to articulate the idea that the Fourier transform might play a role in brain/mind neurophysics. Pribram (1990) spent decades performing laboratory research to gather experimental data in an effort to solve two problems:

(a) to identify the location and mechanism of memory storage (the *engram*), and (b) to discover the cognitive mechanism behind visual perception. Pribram (2013) arrived at the conclusion that the data revealed evidence of Fourier signal transformations of visual signals from the rods and cones of the eyes, and that these Fourier patterns could be detected in spatial Fourier patterns over wide areas of the brain, as fields within the fine-fibered dendritic networks of the cerebral cortex (p. 82).

In the mid-1960s, Pribram was inspired by reports of the first optical holograms, and the empirical evidence that holograms could store, retrieve, and process vast quantities of information using resonant photons in high frequency beams. Ten years later, Pribram (1971) published *Languages of the Brain*, in which he detailed his new theory, the holonomic brain/mind theory, based upon evidence of the Fourier transform playing a key role in the mind/brain process. The theory he put forth proposed that the cognitive sensory processes of memory, sight, hearing, and consciousness in general, may all operate holographically, in a transformational process of information-coded-energies flowing back and forth between space– time and the frequency domain via a Fourier transform mechanism.

Pribram's (1971) theory was radical and controversial, challenging two prominent paradigms of modern neurophysical research: (a) the belief that consciousness is an epiphenomenon produced by electrical sparks among synaptic–clefts throughout the wiring of neurons the brain, and (b) the belief that somewhere in the physical brain, *engrams* of memory are stored, and will be eventually found. Pribram (1971) relates a story of a conversation he had at the time, while climbing with colleagues on a hike in Colorado just prior to attending a neuroscience conference in Boulder:

> We had climbed high into the Rocky Mountains. Coming to rest on a desolate crag, a long meditative silence was suddenly broken by a query from Campbell: "Karl, do you really believe it's a Fourier?" I hesitated, and then replied, "No Fergus, that would be too easy, don't you agree?" Campbell sat silently awhile, then said, "You are right, it's probably not that easy. So what are you going to say tomorrow down there?" I replied, this time without hesitation, "That the transform is a Fourier, of course." Campbell smiled and chortled, "Good for you! So am I." (p. xvii)

Pribram's (1971) hypothesis was strengthened through a growing appreciation of holography as frequency-superpositioned electromagnetic wave interference (p. 142). Pribram called his approach "the holonomic brain theory," and postulated the importance of the *frequency domain* in future research:

> Essentially, the theory reads that the brain at one stage of processing performs its analyses in the frequency domain . . . a solid body of

> evidence has accumulated that the auditory, somatosensory, motor, and visual systems of the brain do in fact process, at one or several stages, input from the senses in the frequency domain. (Pribram, 1982, p. 29)

In Pribram's (1990) theory, a pure frequency domain links with the neuronal tissue of the brain through modulating fields of flux within the fine-fibered dendritic webs of the cerebral cortex regions. His paradigm was reinforced at a San Francisco conference during a lecture given by the physicist Geoffrey F. Chew, the head of the UC Berkeley physics department and a former student of Enrico Fermi. Chew presented a conceptual diagram of the Fourier transform process (fig. 6), which perfectly encapsulated what Pribram had by then become familiar with, the Fourier transform (Pribram, 2004a, p. 230). As shown in the figure, the spectral (frequency) domain, located at the left of the diagram, is directly linked to the spacetime domain, depicted at the right, bridged by the Fourier transform, operating at the sub-atomic levels predicted by Planck's constant.

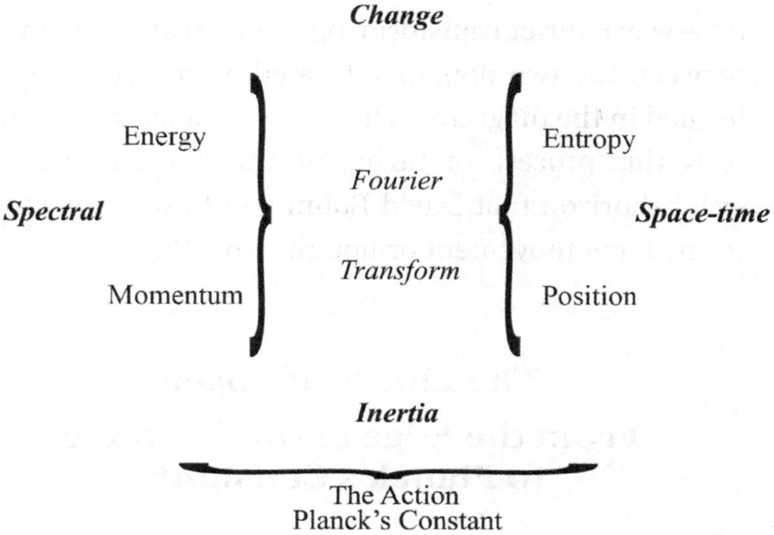

Fig. 6. The Dirac Fourier transform diagram. Source: Adapted from Pribram (2004b)

Pribram (2013) asked Chew where he had obtained the diagram, and was told that he had been given the diagram by his colleague at Berkeley, the physicist Henry Stapp, who himself said he had been given it directly from the British theoretical physicist Paul Dirac (1902–1984), one of the original founders of quantum mechanics. Whatever the origin of the figure, Pribram chose to include the diagram in several future papers. In "Consciousness Reassessed," Pribram's (2004b) caption to the figure reads, "The Fourier Transform as the Mediator between Spectral and Spacetime " (p. 8).

In the diagram, the spectral domain is shown at the left and spacetime to the right, with the Fourier transform between them. The diagram became foundational to Pribram's understanding. It presents a two–way Fourier transform, operational at the boundary between the two domains, located at an event horizon termed in the diagram, "The Action: Planck's Constant." It is this process of turbulent transformation at the event–horizon that David Bohm and Basil Hiley (1993) termed holomovement or holoflux (p. 382).

The Limits of Space

From the Edge of the Universe to Planck's Constant

The Pribram–Bohm hypothesis holds that the dimensions of space are finite and that space exhibits a limited domain in a quantifiable range. This is consistent with the physics of string theory or M–theory, [quote on dimensions moved to first chapter]

Carr (2005) uses the alchemical image of the ouroboros (fig. 7) to illustrate his GUT theory (Grand Unified Theory) in comparing major scale-dependent structural levels of the physical world: "The significance of the head meeting the tail is that the entire Universe was once compressed to a point of infinite density (or, more strictly, the Planck density)" (p. 13). This archetypal figure implies the interconnectedness of the entire universal process in time and space, presenting a cybernetic feedback loop operational at every scale. Mystics have intuited this ouroboric process symbolized in the images of a snake swallowing its own tail (the image has been found as early as the 14th century BCE in the tomb of Tutankhamun) and it is frequently used to symbolize cybernetic feedback in control and communication theory (Wiener, 1948).

While Wiener coined the term *cybernetics*, communication engineers would more commonly see this as metaphor for the "feedback loop," used everywhere in electronic circuit design.

Fig. 7. Alchemical Ouroboros (Pelekanos, 1478)

Stretching out this circular cosmic ouroboric serpent from head to tail, one can create an axis of scales that encompasses all of space. In fig. 8 such a scale is drawn starting with the currently estimated diameter of

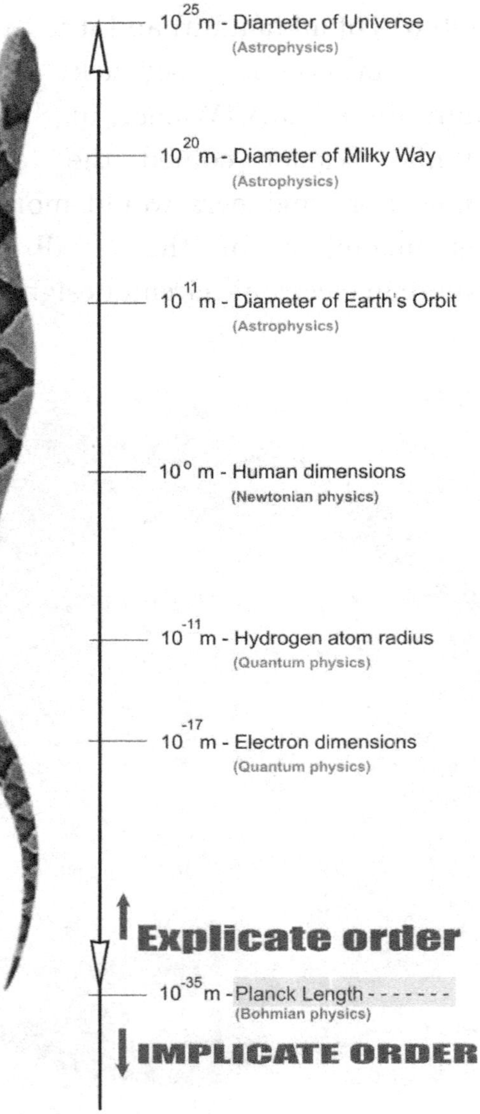

Fig. 8. Scales of dimensional space and the explicate/implicate boundary

the universe itself at 10^{+25} m, and descending logarithmically down to the Planck length limit at 10^{-35} m. The axis thus spans a total range of 10^{+60} (60 jumps by the power of 10). The Pribram–Bohm hypothesis holds that there, at the very bottom of the linear scale (fig. 8), is to be found the transition bounding the explicate order and the implicate order . Here, at the bottom bound of the spatial scale, space reaches its *end*, according to modern physics; but it also marks the *entry point into* Bohm's "implicate order," what Pribram terms the "frequency domain" (Bohm, 1980; Pribram, 2013).

The implications of this topology are profound. Imagine moving inwardly, from any position in the universe, moving into a spherical bubble, shrinking ever smaller in scale while moving ever closer to the center at the bottom of the spatial scale, following the radial axis inward, ever shrinking downward, and then abruptly reaching the end of the line at the Planck length limit of space, the locus of a spherical shell 10^{-35} meters in diameter, below which space has no meaning. Here a boundary has been reached, an event horizon between space and the implicate order. To understand this, one must realize that the classical Cartesian assumption that space is continuous is *wrong*; there *is* indeed a bottom to space, at least according to physics, below which space no longer has meaning. Here there is a discontinuity, as David Bohm and F. David Peat (1987) explain in describing the granularity of space:

> What of the order between two points in space? The Cartesian order holds that space is continuous. Between any two points, no matter how close they lie, occur an infinite of other points. Between any two neighboring

> points in this infinity lies another infinity and so on. This notion of continuity is not compatible with the order of quantum theory Thus the physicist
>
> John Wheeler has suggested that, at very short distances, continuous space begins to break up into a foam-like structure. Thus the "order between" two points moves from the order of continuity to an order of a discontinuous foam. (pp. 311–312)

Pribram's Spectral Flux and the Implicate Order

In 1979, Karl Pribram, at that time a Stanford professor, attended a conference in Cordoba, Spain, where he met David Bohm, a professor of theoretical physics at London University (Cazenave, 1984). During the conference, Pribram (2013) soon realized that David Bohm's model of the implicate order and its projection, or extrusion into space–time, could be seen as entirely compatible with his own holonomic mind/brain theory. Thus began 20 years of correspondence and dialog between David Bohm and Karl Pribram, and the two soon became personal friends.

Pribram (2013) saw in Bohm's theories how the frequency domain flux might be seen to unfold into explicate domain waves of encoded information via the Fourier transform, and he appreciated Bohm's description of how information from the explicate may fold back into the implicate in a bi-directional process. Even more intriguing was Bohm's belief that, "the basic relationship of quantum theory and consciousness is that they have the implicate order in common." (Bohm & Hiley, 1993, pp. 381–382)

Pribram was equally impressed with Bohm's explanation of nonlocality, a major mystery in quantum physics, which Bohm explains as fundamental to the process of folding and unfolding between explicate and implicate orders, allowing for full superpositioned cohesion of frequency information within the implicate order, and even providing a plausible mechanism for Sheldrake's theories of morphogenetic fields and morphic resonance:

> The implicate order can be thought of as a ground beyond time, a totality, out of which each moment is projected into the explicate order. For every moment that is projected out into the explicate there would be another movement in which that moment would be injected or "introjected" back into the implicate order. If you have a large number of repetitions of this process, you'll start to build up a fairly constant component to this series of projection and injection. That is, a fixed disposition would become established. The point is that, via this process, past forms would tend to be repeated or replicated in the present, and that is very similar to what Sheldrake calls a morphogenetic field and morphic resonance. Moreover, such a field would not be located anywhere. When it projects back into the totality (the implicate order), since no space and time are relevant there, all things of a similar nature might get

> connected together or resonate in totality. When the explicate order enfolds into the implicate order, which does not have any space, all places and all times are, we might say, merged, so that what happens in one place will interpenetrate what happens in another place. (Bohm & Weber, 1982, pp. 35–36)

Bohm's topology is both supported and extended by Pribram's contention, supported by the diagram handed down from Dirac, that the boundary or event horizon between the two domains, where the action occurs, is at the Planck length, precisely where, as Pribram tells us here, spectral density in-formation translates into spacetime ex-formation.

> Matter can be seen as an "ex-formation," an externalized (extruded, palpable, compacted) form of flux. By contrast, thinking and its communication (minding) are the consequence of an internalized (neg-entropic) forming of flux, its "in-formation." My claim is that the basis function from which both matter and mind are "formed" is flux (measured as spectral density). (Pribram, 2004b, p. 13)

This flux or spectral density is for Pribram real, in the same sense that spacetime is considered to be real, but this flux is *outside of* or *beyond* space–time. It is in this sense that Pribram made the conceptual leap from considering the Fourier transform as simply a tool of mathematical calculation, to a dawning realization that the reality of the transform implies the ontological *reality* of a domain *outside of space–time*, a transcendent yet ontologically real domain where energy as flux is "measured as spectral density."

Dirac's original diagram can now be extended to include Bohm's two regions of the whole, the implicate order and the explicate order. Figure 9 depicts this expanded diagram.

Fig. 9. Dirac Fourier diagram with David Bohm's topology

An anthropomorphic view of the Pribram's diagram can be seen in fig. 10, where an iris-like lens peering out from the implicate order is maintaining a focus upon and/or projecting a holonomic universe within the explicate order of space–time. This mirrors Karl Pribram's (1991) conceptualization of a lens between the two domains, expressed here in *Brain and Perception*:

> These two domains characterize the input to and output from a lens that performs a Fourier transform. On one side of the transform lies the spacetime order we ordinarily perceive. On the other side lies a distributed enfolded holographic–like order referred to as the frequency or spectral domain. (p. 70)

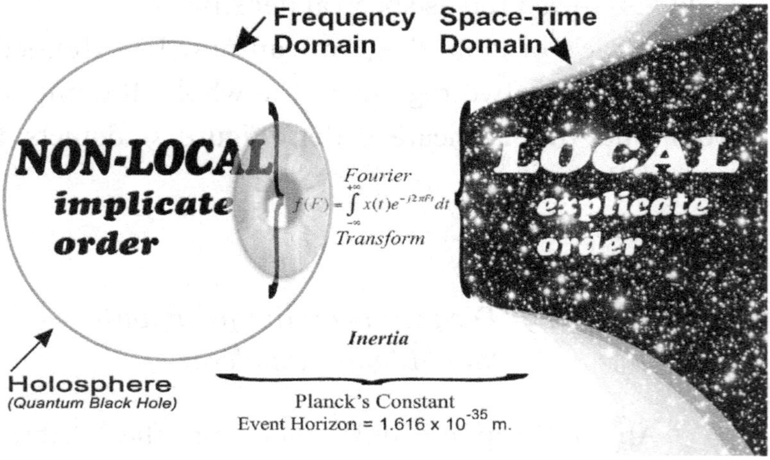

Fig. 10. Topology of consciousness

Note that the image of an iris in the diagram appears at the edge of the event horizon of a quantum black hole, or implicate order holosphere. The iris symbolizes consciousness looking *out* from the implicate order *into* spacetime via a Fourier transform lensing process. This approach to a topology of consciousness as something that is looking out and seeing itself is supported here by the mathematician G. Spencer-Brown (1972) in *Laws of Form*:

> Now the physicist himself, who describes all this, is, in his own account, constructed of it. He is, in short, made of a conglomeration of the very particulars he describes, no more, no less, bound together by and obeying such general laws as he himself has managed to find and record. Thus we cannot escape the fact that the world we know is constructed in order (and thus in such a way as to be able) to see

itself. This is indeed amazing. Not so much in view of what it sees, although this may appear fantastic enough, but in respect of the fact that it can see at all. But in order *to do so, evidently it must first cut itself up into at least one state which sees, and at least one other state which is seen. In this condition it will always partially elude itself. (p. 105)*

Cosmology and the Implicate Order

In 1980 Bohm published *Wholeness and the Implicate Order*, and in a section in which he discusses the cosmology of the implicate order, he puts forth a solution to the problem of "zero-point" energy by regarding the Planck length as the shortest wavelength possible:

> If one were to add up the energies of all the "wave-particle" modes of excitation in any region of space, the result would be infinite, because an infinite number of wavelengths is present. However, there is good reason to suppose that one need not keep on adding the energies corresponding to shorter and shorter wavelengths. There may be a certain shortest possible wavelength, so that the total number of modes of excitation, and therefore the energy, would be finite. When this length is estimated it turns out to be about 10^{-35} m. *(Bohm, 1980, p. 190)*

Bohm (1980) brings up the school of Parmenides and Zeno, which held that all of space is actually a plenum, and he points out that as recently as the last century this same theory was presented in the widely accepted hypothesis of the *ether* (p. 191). Bohm describes how there is a "holomovement" (p. 151) in this immense sea of "zero-point energy" (p. 190) to be understood as a "undivided flowing movement without borders" (p. 172) and he goes on to state:

> It is being suggested here, then, that what we perceive through the senses as empty space is actually the plenum, which is the ground for the existence of everything, including ourselves. The things that appear to our senses are derivative forms and their true meaning can be seen only when we consider the plenum, in which they are generated and sustained, and into which they must ultimately vanish. (p. 192)

In the Pribram–Bohm cosmology then, the interface or boundary between the spacetime explicate domain and the nonlocal, nontemporal implicate domain can be viewed topologically as a holoplenum of holospheres (Figure 8). Here can be found an answer to the "hard problem of consciousness" posed by Chalmers (1995), for it is from *within* each holosphere that consciousness is "peering out" into and "projecting" the spacetime explicate, and here Bohm (1980) summarizes his cosmological essay by proposing that "consciousness is to be comprehended in terms of the implicate order, along with reality as a whole" (p. 196) and stating unequivocally that "the implicate order is also its primary and immediate actuality" (p. 197).

How Spacetime Projects from the Holoplenum

To summarize thus far, the Pribram–Bohm model envisions the existence of a nonlocal, transcendent mode of consciousness out of which projects the spatial cosmos holographically from an infinite plenum or matrix of quantum black holes, each at the limiting Planck length diameter of 10^{-35} m, located at the very bottom of space, at the very center of every three-dimensional spatial coordinate. According to quantum theory, within the implicate order, below the Planck length, space and time do not exist, as Bohm (1980) describes here:

> We come to a certain length at which the measurement of space and time becomes totally indefinable. Beyond this, the whole notion of space and time as we know it would fade out, into something that is at present unspecifiable When this length is estimated it turns out to be about 10^{-33} cm. This is much shorter than anything thus far probed in physical experiments (which have gone down to about 10^{-17} cm or so.*) (p. 190–191)*

This leads to a new vision of the metaverse, a model in which the cosmos can be visualized as projecting outwardly from a holoplenum, from each

Planck holosphere (quantum black hole) at the bottom of space, everywhere. In such a cosmology, the Big Bang theory would need to be revised to encompass more than simply a single point at a single time. The holoflux theory offers the prospect of multiple "Big Bangs" emanating from every individual Planck holosphere, perhaps a continuous recurrence cycling at Planck time (10^{-44} s) quantum ticks. Like images arising from pixels on a two-dimensional LCD (liquid crystal display) screen, the holoplenum projects a three-dimensional cosmos in all its glory. Yet during the process not only is radiant energy projecting *outwardly from* these quantum holospheres, but there is a simultaneous torrent of information, perhaps encoded in the form of gravitationally modulated energy flowing *inwardly into* the implicate order at the center of each and every point in space.

One can only imagine the transitional region between the explicate and implicate order as a frothing, turbulent, resonant event-horizon that bounds spacetime and the transcendent implicate order. Information flowing inward eventually leaves spacetime entropically and flows *into* the network of quantum black holes, the holoplenum of holospheres. Information flowing into this region of infinite centers becomes immediately nonlocal, superpositioned, omni-intersecting. The holoplenum of holospheres provides a vast 3D projection field into which spacetime is projected by implicate order holoflux.

But in addition to this spatial topology, an examination of the transition between time (in the explicate order) and timelessness (in the implicate order) is in order. How might time communicate with the timeless? What happens at the event-horizon of time?

The Digital Universe: Laminated Spacetime

At the heart of quantum theory is the conviction that neither time nor space are continuous and it has been concluded that both are granular at their lowest scale (Smolin, 2012). Having thus far established the granularity of space at the Planck length of 10^{-35} m, what then might be the granularity of time? One approach has been provided by the theoretical physicist John Archibald Wheeler (the originator of the term "black hole"), who, in considering a topology of information in the cosmos, noted that the highest possible clock rate must be limited by the Planck constant in general, and specifically by the *Planck time* value, below which time can have no meaning according to modern physics (Wheeler, 1990). The *Planck time* (t_P) is the time it would take a single photon travelling at the speed of light to cross a distance equal to one Planck length; the Planck length has been determined by the National Institute of Standards and Technology (n.d.) to have a value of 5.39116×10^{-44} s. Wheeler (1990) here describes the granular activity within time and space:

> Space––pure, empty, energy–free space––all the time and everywhere experiences so-called quantum fluctuations at a fantastically small scale of time, of the order of 10^{-44} seconds.

> During these quantum fluctuations, pairs of particles appear for an instant from the emptiness of space. (p. 222)

Given the limit of the smallest interval of time being 10^{-44} seconds, one can calculate the maximum clock–speed of the universe by determining how many Planck time intervals are possible within one human second:

$$(1 \text{ sec} / 10^{-44} \text{ sec}) = 10^{+44}$$

Thus it has been proposed that if the universe is found to operate digitally, it would be cycling at its highest possible clock rate, switching between the explicate order and the implicate order cyclically at a rate of 10^{44} times per second at a bandwidth of 10^{44} Hz. This model has been developed in *The Theory of Laminated Spacetime* by Dewey (1985). The theory accords well with the Pribram–Bohm hypothesis process, and describes a process of energy flux extruding into spacetime as a series of quantum shells of information–encoded energy moving out in quantum jumps at the speed of light, with each quantum shell or brane separated from the next by a spatial gap equal to the Planck length, as described here:

> I believe the universe to be composed of nothing but shells of electromagnetic particles which the theory of Laminated Spacetime describes as laminae of spacetime . (p. 95)

Holograms Versus the Holoplenum

To better understand the projection of the spacetime cosmos from the underlying holoplenum, the distinction between a hologram and a three dimensional holoplenum projection must be drawn. A hologram is an optical image stored and recreated by a single laser (i.e., LASER: Light Amplitude Stimulation of Electromagnetic Radiation), a beam of coherent electromagnetic radiation of a single frequency (Pribram, 1971, pp. 145–152). Human technology greatly simplifies the creation and replay of a hologram by: (a) using a single beam of coherent photon energy (a *single frequency*), and (b) illuminating the object from only two distinct fixed points in space. An interference pattern forms from the beams of intersecting light impinging upon the three–dimensional planes of the object from different angles, causing complex shadows which are detected and recorded on the flat detector plane behind the object (Pribram, 1971, p. 142).

In nature, however, the situation is vastly richer in complexity: electromagnetic flux interactions in spacetime form a highly complex three–dimensional matrix of intersecting shells of every spacetime radiation frequency band conceivable, impinging from an infinity of directions. In contrast with the human generated holograms of a single frequency taken from two fixed points, each point in spacetime actually intercepts the *entire frequency flux spectrum* of the cosmos, which is being continuously captured by and feeding the event horizon entropy of the quantum black hole located at its geometric central "point" (Wheeler, 1998, pp. 313–314).

The omnidirectional interactions manifesting among electromagnetic waves in the dimensional range of the explicate order are mirrored in the spectral frequencies within the implicate order. There is a resonance between implicate and explicate mediated by mathematical relationships such as the Fourier transforms and Gabor functions. This process between the explicate and implicate order is skillfully articulated here by David Bohm:

> The implicate order can be thought of as a ground beyond time, a totality, out of which each moment is projected into the explicate order. For every moment that is projected out into the explicate there would be another movement in which that moment would be injected or "introjected" back into the implicate order. If you have a large number of repetitions of this process, you'll start to build up a fairly constant component to this series of projection and injection. That is, a fixed disposition would become established. The point is that, via this process, past forms would tend to be repeated or replicated in the present, and that is very similar to what Sheldrake calls a morphogenetic field and morphic resonance. Moreover, such a field would not be located anywhere. When it projects back into the totality (the implicate order), since no space and time are relevant there, all things of a similar nature might get connected together or resonate in totality. When the explicate order enfolds into the implicate order, which does not have any space, all places and all times are, we might

> say, merged, so that what happens in one place will interpenetrate what happens in another place. (1981, pp. 35–36)

This resonance between the implicate and the explicate orders nudges into existence a cosmic holonomic metaverse of galaxies, butterflies, and zebras (Sheldrake, 1981).

Geocentric Topology of the Holonomic Metaverse

The topology of the holonomic metaverse is congruent with cosmological intuitions of numerous classical thinkers. In the 4th century BCE, a model proposed by Plato, and further developed by his student Aristotle, consisted of system of numerous crystalline shells rotating about a central sphere located at the center of the Earth (C. G. Fraser, 2006, p. 14). By the 2nd century CE, this geocentric model had become codified by the Alexandrian astronomer Claudius Ptolemy in his cosmological model of concentric spheres (fig. 11).

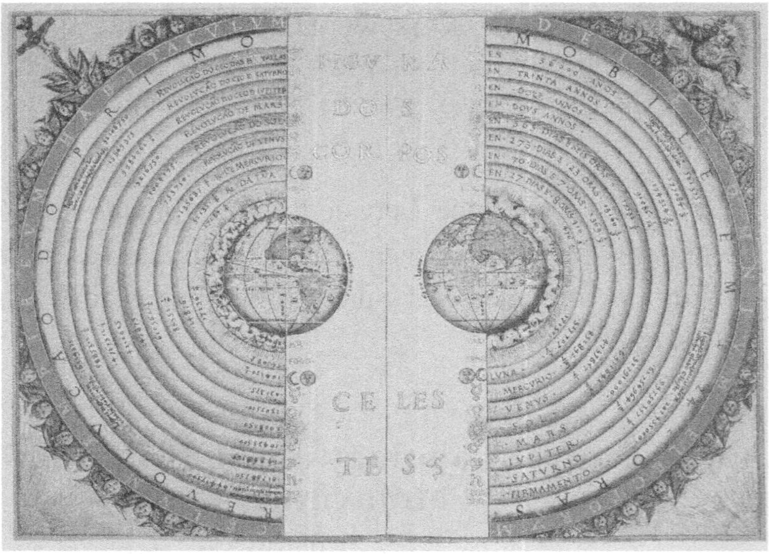

Fig. 11. The Ptolemaic Geocentric conception of the Universe. Source: Velho (1568)

In the 3rd century BCE Aristotle (2004) held that the reason an apple falls to the ground is because it seeks its natural place at the center of the universe, and he set forth a geocentric model based upon the following three propositions:

1. The Earth is positioned at the center of the universe.

2. The Earth is fixed (nonmoving) in relation to the rest of the universe.

3. The Earth is special and unique compared to all other heavenly bodies.

Substituting "holosphere" for "Earth" in Aristotle's propositions, each Planck holosphere can be taken as positioned at the center of the its universe; each holosphere is fixed (nonmoving) in relation to all other holospheres in the holoplenum, and each holosphere is "special" by virtue of its unique Hilbert space coordinates (Young, 1988).

A first approach to a topology of the holoplenum is to begin at the event horizon of the Planck holosphere. How close to the surface of the Planck holosphere does spacetime actually begin? The overriding constraint is that it *cannot* begin any closer than one Planck length, because anything less than that has no meaning in space–time. Accordingly, the first shell, or "isosphere," must be located exactly *one Planck length from* the central holosphere (fig. 12). Moving radially outward from the central holosphere, each subsequent isosphere enclosing the previous isosphere. They can be visualized as nested, like hollow Russian dolls, or spherical tree rings. Within space, each isosphere is separated from the previous by a distance of 10^{-35} meters, the Planck length.

But of what do these isospheres consist? This isospheric shell cannot itself attain any appreciable depth or spatial thickness due to the Planck length constraint; they cannot be within the explicate order of spacetime while their thickness is below the Planck length. Yet if they are thinner than the Planck length, they are too thin to be in space.

It becomes clear that these isospheric shells are *event horizons* between the implicate order (spacetime) and *the implicate order* (the implicate order). Indicated in the figure, between each shell is a sort of dielectric, not space, but the implicate order. but a quantum vacuum which may also be the locus of Bohm's (1980) "infinite zero-point fluctuations of the 'vacuum field'" (p. 70).

A topology of the implicate order is here coming into focus. A nonlocal, timeless, transcendent domain is to be found at the center, everywhere, in the form of Planck diameter holospheres making up an implicate holoplenum, and enclosing each and every holosphere, nested at quantum multiples from the Planck length to the diameter of the universe, are isospheric shells of the implicate order.

While distributed geometrically within the explicate order, these isospheric quantum shells are not in space, but share in the nonlocality of the implicate order. They provide both a fundamental structure and information backbone bridging the implicate order and the explicate orders. And each isosphere provides a structural capacity for information storage limited only by the Bekenstein bound (discussed in the next section).

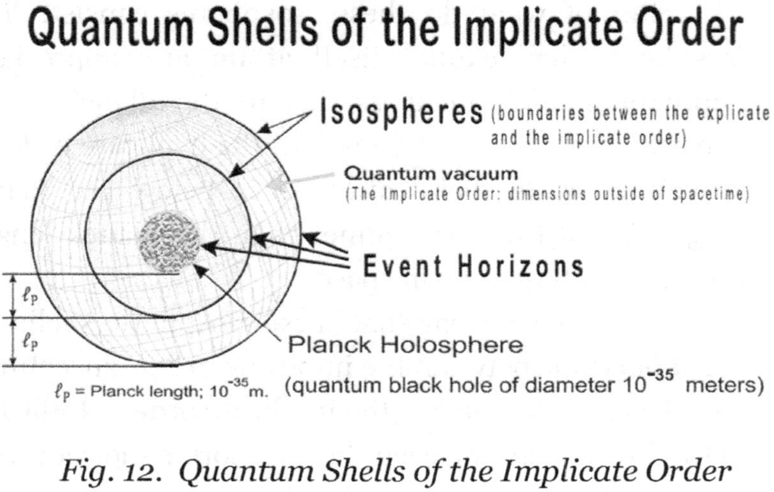

Fig. 12. Quantum Shells of the Implicate Order

As in all black hole phenomena, these isospheres (as well as the central Planck holosphere) are assumed to have an angular momentum (spin) due to turbulence at the event horizon, a highly energetic region which, the holoflux theory posits, are modulated with qubits of information. These spinning, implicate order isospheres project spacetime "objects" (multidimensional holograms) from frequency-phase information within the implicate order.

Spectral frequencies within the implicate order (the frequency domain of communication engineering) can be correlated with sets of isospheres having explicate order diameters corresponding to wavelengths of the same frequency. Thus a complex patterned frequency spectrum within the implicate order can be mirrored in the spatial dimension of the explicate order through the activation of selected ranges of shell isospheres of the implicate.

Having established the mechanism of communication linking Bohm's (1980) implicate and explicate orders, it is now possible to turn to memory storage, that elusive problem which first led Pribram (1971) to his holonomic theory of a brain/mind interface.

A Holoplenum of Quantum Black Holes

As described in the previous section, a holonomic process underlies the Bohmian "Whole" and operates through means of a cosmic plenum of quantum black holes. These quantum black holes present as geometric shells or spheres in space, each one being a boundary between space (outside of the sphere) and the hidden dimensions of the implicate order within the shell boundary (though "within" is a spatial metaphor). Each of these quantum black holes is exactly one quantum in diameter, that is, one Planck length. Thus these quantum-diameter shells have been termed "Planck diameter holospheres."

In geometric terms, these "close-packed spheres" refers to the most tightly packed and highly efficient arrangement of spheres of equal diameter, each of which touches multiple similar adjacent spheres. The entire geometric plenum can be visualized at the very bottom of space, existing everywhere in our spacetime universe, and considered metaphorically as a three-dimensional "ocean floor" of space.

How Holopixels Project the Universe

These inconceivably tiny holospheres act as pixels, or *holopixels*, continually projecting outward into our four dimensional universe of spacetime from the hidden dimensions of the metaverse. Much in the same way that the two-dimensional array of pixels on the back of a plasma display screen projects a streaming video image that we can see, this plenum of quantum holopixels projects the universe into our space and time dimensions. Yet, the region within the boundaries of these black holes contains the multiple hidden dimensions beyond spacetime .

This three-dimensional substrate or plenum exists everywhere and continually projects our spacetime universe that we call the cosmos (fig. 13).

Fig. 13. A Seer at the Intersection of Heaven and Spacetime[5]

In totality, this plenum of holopixels acts much as the two-dimensional plenum of pixels that are embedded in the substrate of a modern large screen plasma video display. Surrounding and radiating outward from each central Planck holosphere throughout space are spherical shells of quantum potential, infinitely thin information-encoded shells of energy, each separated from an inner shell by the quantum exclusion radial distance of one Planck length. These isospheric shells of Bohmian quantum potential extend out to the diameter of the universe itself.

Thus can be visualized topologically an almost infinite series of nested isospheric shells, spread out, at discrete quantum radii from their respective central

Planck holospheres, each bounding Bohm's nondual implicate order.

The global panoramic intersection of these holospheric shells manifests as the projection of a the holonomic universe into spacetime . The cumulative effect of this projection, as regarded by human physicists observing from significantly higher scalar dimensions, is described as "matter." The phenomenon can be understood as a *process of projected creation*, an omnipresent, on-going holographic extrusion of information *from* the implicate order *into* the explicate order, where the various structures of the cosmos (galactic clusters, stars, etc.), the complex unfoldings *into* spacetime , are perceived by the human eye and mind to be three-dimensional, when they are actually holographic projections *from* the implicate order, from the center outward.

Another way of visualizing the projected illusion of spacetime reality from the holoplenum is by expanding upon the metaphor of a flat-panel plasma display (such as the one you may be viewing as you read this).

Consider the human visual threshold for detecting separate images, which lies somewhere between 10 to 12 images per second; the industry standard in the motion picture industry is 24 frames per second (Dorf, 1997, p. 1538). This standard ensures that the presentation of a sequence of projected images will appear to a human viewer as a smooth and continuous motion.

By contrast, if the entire universe flashes in and out of existence at a clock-cycle rate limited only by the Planck time constant of 5.3×10^{-44} seconds, equivalent to a “frame” rate of almost 10^{44} “frames per second,” the cosmos would *appear* to be smooth and continuous in all respects even to an electron, and certainly to any human observer of the cosmos, even at quantum dimensions.

The approximate image resolution of a “holoplenum display” obtained by dividing one inch by the Planck length, yields a maximum resolution of 1.584 x 10^{34} holopixels per inch. At such hyper-fine resolution, even a Higgs boson in the 10^{-17} meter dimensional range would appear to be moving smoothly through space.

Holonomic Storage: The Bekenstein Bound

In *Wholeness and the Implicate Order*, Bohm (1980) articulates and develops a "quantum potential" (p. 77) function that projects the explicate spacetime universe out from within an enfolded sub-quantum implicate order. Bohm's quantum potential function is congruent with de Broglie's "Pilot wave" theory of 1927, as both are based upon a conviction that there exist "hidden variables" in sub-quantum regions not accessible to observational exploration using current material science technology (and far beyond the capabilities of the CERN Large Hadron Collider).

The de Broglie pilot wave theory and Bohm's (1980) quantum potential are mathematical attempts to map sub-quantum effects issuing from an implicate order in a domain of "hidden variables" far below the observational capabilities of contemporary material science.

Both theories posit a cybernetic processing of information, simultaneously being cycled from the spacetime world and enfolded into the nondual frequency domain where the accumulating information is processed nonlocally within the implicate order. Driven then by the implicate order, a pilot wave of quantum potential nudges the configurations in spacetime into an altered, slightly new configuration, much as a small tugboat might influence an enormous freighter. If the cosmos operates at its maximum possible clock cycle, as discussed previously, this pilot wave might be seen to operate at the extreme clock cycle rate of the Planck time constant, or 10^{44} Hz.

Somewhere, however, such a cosmic process would require a memory storage repository in spacetime . Regarded as a cybernetic process, the sequence of information feedback and action can be metaphorically imaged in the alchemical Ouroboros, the classical symbol for consciousness, depicted as a snake in a circular configuration eating (or chasing) its own tail. This process can be viewed as the cyclic transfer of information coming in (from the tail) and the resulting action (by the head). A cybernetic feedback loop thus needs data, information, as input. Where then might data be accumulated and retrieved in spacetime at these most fundamental sub-quantum levels, in a Bohmian holonomic universe consisting topologically of the distributed plenum of Planck holospheres, each surrounded by a series of nested quantum isospheres?

One possibility is to consider the information storage potential of an isosphere encoded with granular "bits" of data. In 1970, Jacob D. Bekenstein, then a graduate student working under John Archibald Wheeler (who himself coined the term "black hole"), proposed a novel idea. Bekenstein proposed that there must be an absolute maximum amount of information that can be stored in a finite region of space, and that the Planck constants in quantum theory can be used to determine this limit (Wheeler, 1990). Twenty years later, Bekenstein's theory was extended into what is called the *holographic principle* by Leonard Susskind (1995), which describes how information within any volume of space can be encoded on a boundary of the region. A description of this configuration is presented here by Wheeler (1990) himself, as first related to him by Bekenstein:

> One unit of entropy (information), one unit of randomness, one unit of disorder, Bekenstein explained to me, must be associated with a bit of area of this order of magnitude (a Planck length square). . . . Thus one unit of entropy is associated with each 1.04×10^{-69} square meters of the horizon of a black hole. (p. 222)

This proposed upper limit to the information that can be contained upon the surface of a specific, finite volume of space has come to be known as the "Bekenstein bound" (Bekenstein, 1973). Symbolically depicted in Fig. 14 is a topological depiction of the arrangement of information bits, or "qubits," stored on the bounding surface of a spherical volume or isosphere.

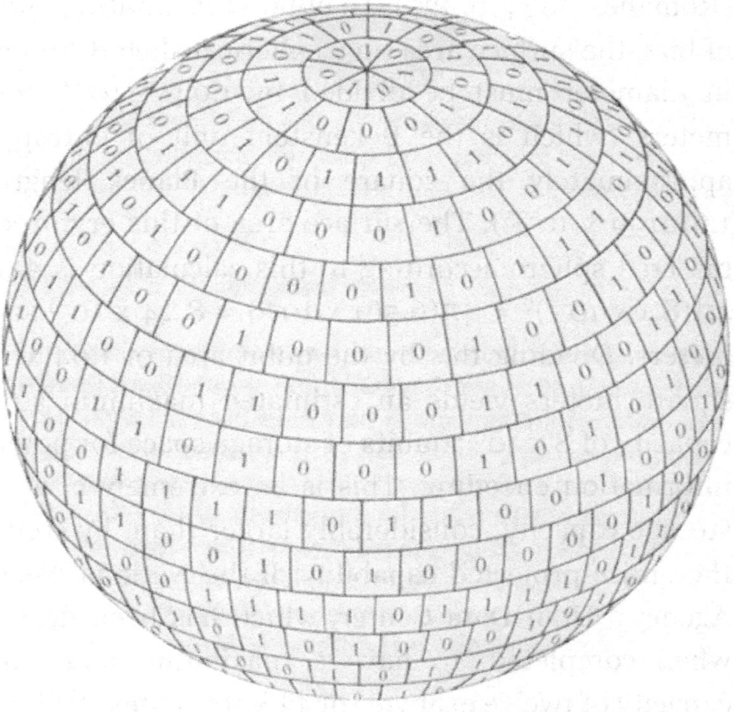

Fig. 14. Planck length qubits on the surface of an isosphere.[6]

This same topological approach to data storage can be applied to human physiology. Using a well-known biological structure as an example, it is possible to calculate the maximum memory storage capacity of a isosphere the size of a single erythrocyte, the ubiquitous red blood cell found throughout the human body. Using Wheeler's approach to determine the number of Bekensteinian equivalent data bits (qubits) on the surface of an isosphere, and using the average diameter of a typical human erythrocyte of 8.1 microns (or 8.1 x 10^{-5} m), the maximum possible storage capacity on the surface of a single red blood cell can be calculated

(Romanes, 1964, p. 137). To obtain this limiting number of bits, the surface area on a spherical shell 8.1 microns in diameter must be divided by 1.04 x 10^{-69} square meters (which is the Bekenstein unit of entropy, or approximately the square of the Planck length of 1.616199 x 10^{-35}). The surface area of this erythrocyte-bisected sphere according to this calculation is $4\pi r^2$or $4\pi(8.1 \times 10^{-5})^2 = 4\pi(6.561 \times 10^{-9}) = 8.24 \times 10^{-9}$ square meters. Dividing this by the qubit area of 1.04 x 10^{-69} square meters yields an estimated maximum storage capacity of 8 x 10^{60} qubits of storage space for potential information encoding. This is an extremely large data storage capacity, considerably larger than, by contrast, the entire projected capability of the National Security Agency's Utah Data Center, which has been designed, when completed, to have a maximum data storage capacity of twelve exabytes or 12 x 10^{18} bytes (Hill, 2013, para. 7)

Unfolding an Implicate into Spacetime

The ontological understanding of quantum physics that Bohm sought emerges in this model of a sub-quantum, omni-present holoplenum of Planck holospheres, each enclosing a contiguous transcendent region of non-spatial, non-temporal dimensions termed by Bohm "the implicate order" (Joye, 2016).

What are the implications of this model for human consciousness, cognitively operational at temporal and spatial scales vastly larger than those found at these sub-quantum Planck boundaries? To answer this we must first complete the Pribram-Bohm cosmological topology of consciousness, and to do this the concept of isospheres, shells within shells, must be considered. Moving outwardly, radially, from the interior bounding event horizon at each central Planck holosphere, can be identified isospheric shells of the implicate order, extruding into space at exact Planck length (quantum) intervals.

This series of concentric shells, each one separated from the next by one Planck length, are isospherical loci of the implicate order (see fig. 5); they extrude into space and they intersect in space with other shells bounding other Planck holospheres in the spacetime holoplenum. It is the cumulative interference effect of the intersection of individual isospheric shells which project images at higher scales, holographically, into three dimensional space.

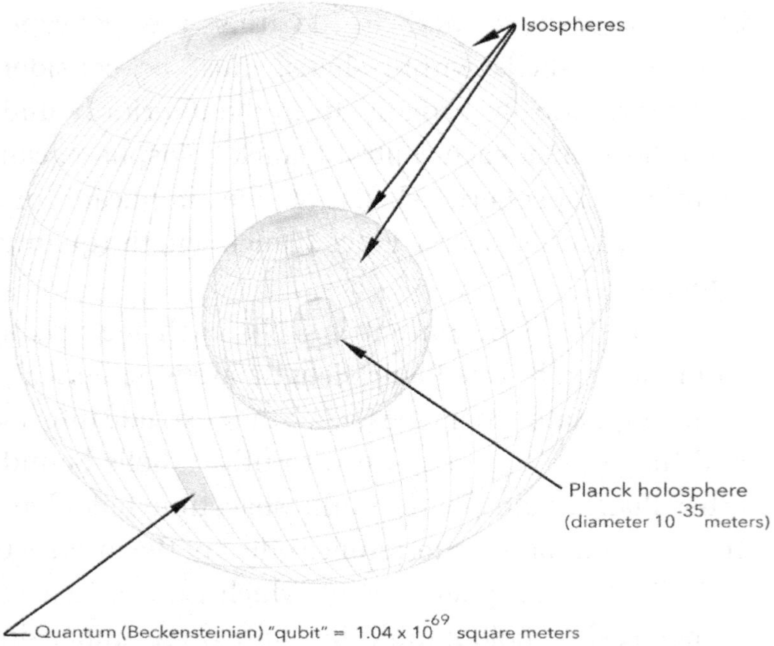

Fig. 15. Topology model of isospheres surrounding a Planck holosphere. Author's figure.

Each isospheric shell, as Bekenstein (1973) determined, has a potentially enormous storage capacity in "qubits" of information encoded on the event-horizon bounding each shell, depending only upon the radius of the shell within the range of 10^{-35} m to 10^{27} m. Accessible simultaneously in both the implicate order and the explicate order, such encoded information provides the data to guide evolving forms as they project into the explicate via the pilot wave mathematics of Bohm's (1980) "many-dimensioned quantum potential" (p. 80). As part of this process, in-formation becomes ex-formation as the implicate order unfurls into the explicate. The plasmoidal forms appearing in spacetime as electromagnetic flux energy are mirrored by and resonate within the implicate order as the dark energy of frequency-phase holoflux.

Within this cosmic geometry of the Pribram-Bohm hypothesis can be identified a framework for omnipresent two-way portals, potential bridges between the explicate and the implicate. Here, the deep consciousness of the universe flows in a cyclic, cybernetic, perhaps fractal movement in time and space, moving through an endless, bi-directional process, consciousness involving itself in a dance of transformation.

How then does this topology support human consciousness, thought, and perception in spacetime ? How can electromagnetic frequency plasma in spacetime resonate with holoflux plasma in the implicate order? First of all, the energies must be within the same frequency range in order for maximum interactive resonance to occur. Where the frequencies overlap as they superimpose and interpenetrate one other, resonance occurs.

Resonance is a naturally occurring phenomena characteristic of physical objects or plasma fields extended in spacetime . The resonance effect is seen when objects or complex signal systems exhibit remarkable frequency sensitivity to particular external frequencies, flowing in, through, and around the system, frequencies which approach the "natural resonant frequency" of the object or signal system; perfect resonance occurs where the input frequency and the natural frequency are identical (Feynman, Leighton, & Sands, Leighton, & Sands, 1964, p. 78). This principle of resonance governs all cybernetic feedback loops, and is a key factor in the design of antennas for electromagnetic transmitting and receiving systems. The goal of antenna design is to construct an antenna that is maximally resonant within a specific narrow frequency range of incoming (external or internal) electromagnetic radiation; when the input frequency and the natural frequency of the antenna coincide or move significantly close to one another, resonance occurs. This simplest and most common antenna design is the dipole antenna, which depends directly on the size (wavelength) of the incoming electromagnetic wave. A dipole antenna is designed to be physically half the size of the incoming wavelength, as stated in a textbook of antenna design:

> A fundamental form of an antenna: length is approximately equal to half the transmitting wavelength. It is the unit from which many more complex forms of antennas are constructed. It is known as a dipole antenna. (Blakeslee, 1972, p. 580)

This same half-wavelength effect governs the design of network communication waveguides applied to fiber-optics in the Internet, where the antenna is the fiber channel itself acting as a waveguide. The waveguide is highly efficient for two reasons: first, as its name implies, the waveguide guides the electromagnetic wave within its channel with maximum efficiency; and secondly, it shields the signal in the channel from external electromagnetic waves. The inner diameter of the hollow waveguide is designed to be equal to exactly half the wavelength of the electromagnetic energy signal shielded by and flowing through the waveguide channel.

Waveguides have been used for over a century both commercially and in research to channel and guide vibrating energy of specific limited frequency ranges; the fiber-optic networks hosting the global internet operate on this principle, channeling electromagnetic radiation at fixed laser frequencies (Dorf, 1997). It was discovered late in the 19th century that circular metallic tubes, or hollow metal ducts, similar to A/C ventilation ducts but much smaller, could be used to channel and guide either sound vibrations in air, or electromagnetic energy in air or vacuum. Without the waveguide, the vibrational energy field is transmitted in all directions, visualized as magnetic lines or arrows emerging from a point at the center of an expanding sphere. This energy disperses outwardly, the magnetic vectored arrowheads pushing into the inside of an infinitely expanding sphere. A waveguide, however, constrains the magnetic component of the wave-front of vibrating energy to one specific linear direction, in parallel with the center of the waveguide, and thus, conceptually, the confined wave itself loses very little power while it propagates along the central axis of the waveguide, like a stream of water emerging from the pinprick of a large, taut, water balloon (Dorf, 1997).

The most common type of fiber-optic cable used in the Internet has a core diameter of 8-10 micrometers and is designed for use in the near infrared (Gowar, 1993, p. 64). The electromagnetic signal wavelength that runs through the global fiberoptic network is powered by highly efficient carbon dioxide lasers, and has a wavelength of 10 micrometers. Coincidentally (or not?), the average human blood capillary diameter is also 10 microns, and blood capillaries are at all times full of carbon dioxide.

Dimensional analysis and a cursory examination of human physiology would immediately suggest two candidates for waveguide systems within the human body: (a) the blood capillary system, and (b) the microtubule system. The corresponding resonant frequency for electromagnetic waves using such waveguides correspond to wavelengths matching the inner diameter of these structures. For blood system capillaries, this corresponds to radiation with a wavelength of 9.3 to 10.0 microns, the average inner diameter of a capillary. For microtubules, the radiation wavelength would be found in a range of 40 nanometers, the inner diameter of the microtubule waveguides. Fig. 16 depicts the location of each of these potential waveguide frequency bands within a wider section of the electromagnetic spectrum.

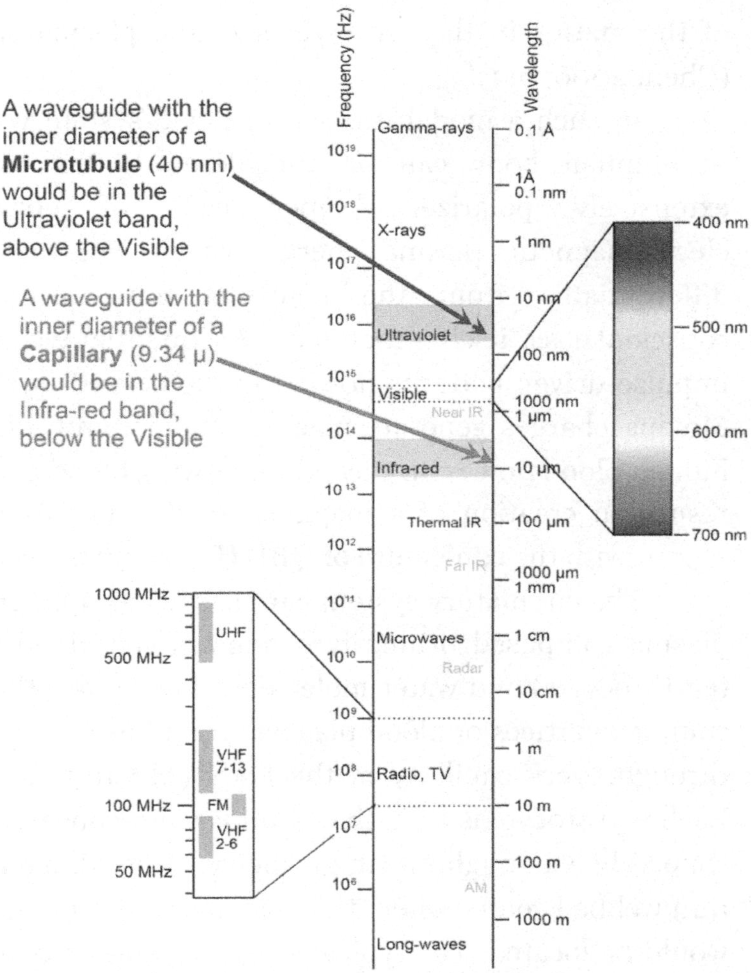

Fig. 16. Microtubules and capillaries as waveguides. Annotations by author; graphic by Jahoe (2012). Reprinted under the terms of a Creative Commons Attribution ShareAlike 2.5 Generic license. Image retrieved from Wikimedia Commons.

There is no impediment to our blood acting as an electromagnetic plasma within the capillary system, and as previously mentioned, the opening page of a textbook on plasma physics reads, "It has often been said that 99%

of the matter in the universe is in the plasma state" (Chen, 2006, p. 1).

In such a model, the entire blood system within the human body can be considered to act as an extensively polarized "super cell" of nonlocal electromagnetic plasma energy, which can then be differentiated from the neuronal brain body of consciousness, itself generated by sequential electrical impulse-driven patterns flowing in the nervous system. Moving charges generate magnetic fields, and ionized human blood flow is no exception: flowing blood plasma results in creation of a magnetic field, and this is in accord with the conjecture of QBD (Dorf, 1997, p. 27).

The circulatory system can be seen as a magnetic plasma composed primarily of ionized red blood cells (erythrocytes) and water molecules, flowing together in complex vortices of blood plasma around every cell and through every capillary of the body (McCraty, 2003). Each erythrocyte is a flexible, annular, bio-concave disk shaped like a doughnut (in geometry, a torus), having a thin webbed center where the hole in a pastry doughnut would be located. The typical outside diameter of a red blood cell is approximately 9 microns, close to the infrared wavelength of 9.6 microns generated by the human body (Turgeon, 2012). The adult human body contains approximately 6 grams of iron, of which 60% is stored throughout the 10^{12} erythrocytes, each of which contains approximately 270 million atoms of ionic iron embedded within transparent hemoglobin in a toroidal locus (Romanes, 1964). Thus each erythrocyte, replete with iron ions embedded in hemoglobin, creates in effect, an ionized iron toroid (Wick, Pinggera, & Lehmann, 2000, p. 6).

Recent studies have also discovered neuronal generation of electromagnetic energy in the near infrared region of the spectrum centered around 10 microns. Radiation emission was repeatedly measured emanating from live crab neurons in extremely narrow, discrete spectral bands within the frequency range corresponding to a spectral region from 10.5 to 6.5 microns (A. Fraser & Frey, 1968).

The implications of this model are considerable: there may exist in nature a unique resonant frequency for each individual human being. It is useful here to step through a topological analysis of the possible functions of a human red blood cell, given its geometry, as a locus of consciousness, and the possible use of the erythrocyte as a locus of memory storage at human biological scales. If, as previously conjectured, the red blood cell has an ideal diameter to resonate electromagnetic radiation in its ferrite-embedded ring at the human infrared wavelength of 10 microns, then it is reasonable to ask if this configuration could accommodate a single unique isospheric frequency (wavelength) for each of the currently 7 billion living humans on the planet. In other words, does this geometry allow for the possibility of each human being to also have a single unique frequency within the infrared electromagnetic radiation that resonates within the human cardiovascular waveguide system? Fig. 7 outlines the topological feasibility of this approach.

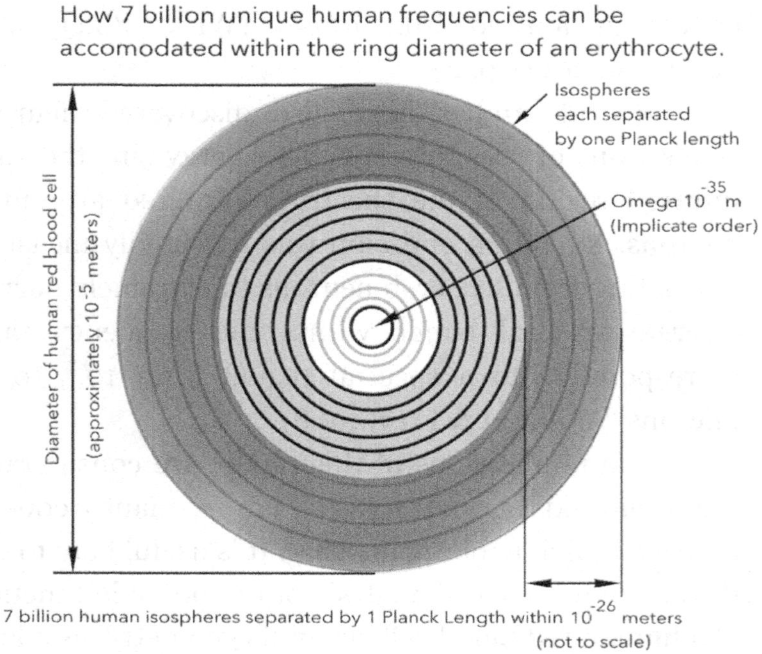

Fig. 17. Isospheric capacity of a single erythrocyte. Author's figure.

Assuming each unique frequency would match its radially unique isosphere, separated by only one Planck length, Figure 17 suggests how 7 billion unique isospheres, each of quantum discrete frequency, might be nested within the geometry of a typical human red blood cell (in the image, a multiple of 7 billion times the Planck length of approximately 10^{-35} m results in an estimated shell thickness of 10^{-26} m). This model supports the feasibility that each living human being might have a unique holospheric frequency, detectable by other human blood cells via the implicate order about which each is centered, and thus provides a possible mechanism for communication, via the mechanisms of resonance, nonlocality, and superposition in the frequency domain of the implicate order.

Sound, Frequency, and the Implicate Order

In contrast with the lack of success others were having in their efforts to establish the mechanism and location of memory engrams, Pribram put forth a theory that the brain mechanics of vision might be seen as holographic processes, projections onto the cortical surface from a frequency domain through the mechanism of frequency superposition and electromagnetic wave interference.[7] Pribram called this "the holonomic brain theory," and postulated the importance of the *frequency domain*:

> Essentially, the theory reads that the brain at one stage of processing performs its analyses in the *frequency domain* . . . a solid body of evidence has accumulated that the auditory, somatosensory, motor, and visual systems of the brain do in fact process, at one or several stages, input from the senses in the *frequency domain*.[8]

Pribram's theory holds that the mathematics of the Fourier transform operates within the brain, modulating electromagnetic fields of flux to create, store, replay, and process holograms within some as yet unidentified physiological context or substrate, but one that Pribram conjectures may be found within the fine-fibered dendritic webs of cerebral cortex regions.[9]

Pribram here takes exception to Bohm's use of the words "flow" or "movement," holding that such words cannot be used to characterize dynamism in a dimension devoid of space or time axes:

> David Bohm (1973) had a concept similar to flux in mind, which he called holomovement. He felt that my use of the term "flux" had connotations for him that he did not want to buy into. I, on the other hand, felt holomovement to be vague in the sense of asking, "what is moving?"[10]

Pribram clearly distinguishes between observations made in the frequency domain and those made in space and time. While the ontological realization of flux in spacetime can be conceptualized in the gyrations of three-dimensional electromagnetic waves, Pribram defines the ontological substrate of spectra in the frequency domain to be pure spectra or holoflux:

> The spectral domain is characteristically a *flux*, composed of oscillations, fluctuations, whereas interference patterns among waves intersect to reinforce or cancel. Holograms are examples of spectra, that is, of a *holoflux*.[11]

This holoflux is essentially the same concept as David Bohm's *holomovement*, though in "Brain and Mathematics,"

Two domains, the time domain (t_d) and the frequency domain (f_d), are of essential importance in the field of information theory and signal communication. On the first page of an undergraduate electrical engineering textbook on network analysis and synthesis, Kuo states:

> In describing signals, we use the two universal languages of electrical engineering—*time* and *frequency*. Strictly speaking, a signal is a function of time. However, the signal can be described equally well in terms of *spectral* or *frequency* information. As between any two languages, such as French and German, translation is needed to render information given in one language comprehensible in the other. Between time and frequency, the translation is effected by the *Fourier series* and the *Fourier integral.*[12]

In his book *Brain and Perception*, Pribram used almost the same words as Kuo in discussing methods by which a Fourier transform operates in mathematical physics. Here Pribram reaches the same conclusion as Kuo:

> It is reasonable to ask: What advantage does the organism gain by processing in the spectral transform domain? The answer is efficiency: The fact that correlations are *so easily achieved* by first convolving signals in the spectral domain and then, inverse transforming them into the spacetime domain. Thus Fast Fourier Transform (FFT) procedures have become the basis of computerized tomography, the CT scans used by hospitals.[13]

The mathematics expressed in the Fourier transform can also be understood as a link between a physics of spacetime (t_d) and an actual frequency domain (f_d). The Fourier and inverse Fourier transforms are shown in fig. 18:

$$f(t) = \int_{-\infty}^{+\infty} X(F) e^{j2\pi Ft} dF \qquad f(F) = \int_{-\infty}^{+\infty} x(t) e^{-j2\pi Ft} dt$$

Fourier integral transform of a continuous frequency function into the *time domain* (t_d).

Fourier integral transform of a continuous time function into the *frequency domain* (F_d).

Fig. 18. Fourier Transform/Inverse Transform.[14]

These Fourier transform expressions indicate that any arbitrary function in the timespace domain, $f(t)$, can be transformed into and expressed by an infinite series of frequency spectra functions $X(F)$ in the frequency domain, and conversely, that any arbitrary function in the frequency domain, $f(F)$ can be transformed into and expressed by an infinite series of time functions, $x(t)$.

These Fourier transforms are themselves derived from an underlying series of alternate pure sine and pure cosine waves, as depicted in fig. 9.

$$f(t) = a_0 + \sum_{n=1}^{\infty} \left(a_n \cos \frac{n\pi t}{L} + b_n \sin \frac{n\pi t}{L} \right)$$

Fig. 19. The Fourier Series.[15]

A century after Fourier's death, Norbert Wiener made use of Fourier's transform to model and analyze brain waves, and he was able to detect frequencies, centered within different spatial locations on the cortex, that exhibited auto-correlation. Specific frequencies were found to be attracting one another toward an intermediate frequency, thus exhibiting resonance or "self tuning" within a narrow range of the frequency domain (f_d).[16] This discovery led Wiener to conjecture that the infrared band of electromagnetic flux may be the loci of "self-organizing systems."[17]

> We thus see that a nonlinear interaction causing the attraction of frequency can generate a *self-organizing system*, as it does in the case of the brain waves we have discussed . . . . This possibility of self-organization is by no means limited to the very low frequency of these two phenomena. Consider self-organizing systems at the frequency level, say, of infrared light.[18]

Wiener goes on to discuss such possibilities in biology, where he focuses upon the problems of communication at molecular and primitive cellular levels, specifically on the problem of how substances produce cancer by reproducing themselves to mimic pre-existing normal local cells. Molecules do not simply pass notes to one another, and they do not have eyes, so how do they perceive and how do they communicate? Wiener conjectures:

> The usual explanation given is that one molecule of these substances acts as a template according to which the constituent's smaller molecules lay themselves down and

> unite into a similar macromolecule. However, an entirely possible way of describing such forces is that the active bearer of the specificity of a molecule may lie *in the frequency pattern of its molecular radiation*, an important part of which may lie *in infrared electromagnetic frequencies* or even lower. It is quite possible that this phenomenon may be regarded as a sort of attractive interaction of frequency.[19]

Perhaps the most fascinating tool for exploring the nature of the interface between the Real–Imaginary domains was developed in 1980 when the Polish–American engineer, Benoit Mandelbrot, created the software to plot an actual image of the two-dimensional interface between spacetime and frequency domains close to the origin (defined as the intersecting point where the Real axis equals zero and the Imaginary axis equals zero). His initial impression, upon seeing the first image, was that the computer program had malfunctioned.[20] Subsequent computer plots assured him that these visual patterns were truly there. Images of this region about the time frequency origin have gained interest worldwide and the region itself has come to be known as the Mandelbrot set (fig. 50). The English mathematical physicist Sir Roger Penrose was so taken by the resulting images that he described them with a sense of almost reverential awe:

> The Mandelbrot set is not an invention of the human mind: it was a discovery. Just like Mount Everest, the Mandelbrot set is just *there*![21]

Fig. 20. Mandelbrot: Origin of real–imaginary axis. Graphic by author.

The Mandelbrot set exhibits remarkable properties. As calculations are done on ever smaller regions on the time–frequency plane, the images appear similar but never completely repeatable, and the viewer begins to sense some sort of biological shapes emerging from this strange world of pure mathematical being. Penrose goes on to say, "The very system of complex numbers has a profound and timeless reality which goes quite beyond the mental constructions of any particular mathematician."[22]

Jung observes the relationship of psyche to matter with the same pattern, noting that they are two different modes of one and the same thing:

> Since psyche and matter are contained in one and the same world, and moreover are in continuous contact with one another and ultimately rest on irrepresentable, transcendental factors, it is not only possible, but also fairly probable, even, that psyche and matter are two different aspects of one and the same thing.[23]

Holoflux, as a total manifestation, is a dual process of reflective consciousness and transcendent awareness. Communication, and consciousness flow between and within the two domains: the implicate and the explicate. Another way of regarding holoflux is to see it as a two-way moving flux: (1) It is an explicate energy (E) moving *inward* into the implicate order at the holospheric event horizon of a quantum black hole, and (2) Moving in the opposite direction, it is an implicate energy (Q) moving *outward* as quantum potential from the event horizon, as holoflux energy transforms from out of the implicate order into the explicate order. My thesis rests upon the assumptions that Bohm's implicate order is real, that the frequency domain is real, and that his active information wavefunction is real, affecting our world of "particles" as it springs from the virtually infinite nonlocal information holoflux within the implicate order.

Like Bohm, electrical engineers separate the world into two domains, which they term the *complex-frequency domain* and the *time domain*. It is our normally perceived spacetime that electrical engineers identify as the *time domain*, but it is in the *frequency domain* that they see the operation of communication signals.[24]

Franklin F. Kuo, an American computer scientist and communications engineer at Bell Laboratories, wrote an electrical engineering textbook, developing the mathematics of communication in *Network Analysis and Synthesis*. In the first chapter of his classic textbook, Kuo describes the two domains, and the mathematical advantage of creating and receiving communication signals within the *frequency* domain as opposed to the *spacetime* domain:

> In the *complex-frequency domain*, the voltage–current relationships for the elements are expressed in *algebraic* equations. Algebraic equations are, in most cases, more easily solved than differential equations. Herein lies the raison *d'être* for describing signals and networks in the *frequency domain* as well as in the *time domain*.[25]

Franklin F. Kuo, an American computer scientist and communications engineer at Bell Laboratories, wrote an electrical engineering textbook, developing the mathematics of communication in *Network Analysis and Synthesis*. In the first chapter of his classic textbook, Kuo describes the two domains, and the mathematical advantage of creating and receiving communication signals within the *frequency* domain as opposed to the *spacetime* domain:

> In the *complex-frequency domain*, the voltage–current relationships for the elements are expressed in *algebraic* equations. Algebraic equations are, in most cases, more easily solved than differential equations. Herein lies the raison *d'être* for describing signals and networks in the *frequency domain* as well as in the *time domain.*[26]

These two orders of electrical engineering theory can be seen to reflect David Bohm's explicate and implicate orders. Bohm's explicate order is associated with the engineering time dimension through which signals are transmitted and received sequentially in time, while Bohm's implicate order can be identified with the electrical engineering *frequency* dimension as a timeless, spaceless domain of pure frequency-phase information.

It has been discovered that these two domains are strongly linked, mathematically, with the operation of the Fourier transform. As Kuo says, "between time and frequency, the translation is effected by the Fourier series."[27] As we have seen in Chapter 4, the Fourier series has also become the bedrock of quantum theoretical calculations, as expressed by David Bohm in the opening of his textbook, *Quantum Theory*:

> It seems impossible to develop quantum concepts extensively without Fourier analysis.[28]

Our thesis, then, is that Kuo's frequency domain, bridged to time by Fourier's mathematics in our twenty-first century information technology, has an ontological reality, and that there does exist an actual frequency domain, which we also identify with Bohm's hypothesized implicate order.

AI, Human, and non-Human Consciousness

In a particularly consequential statement, Bohm, the quantum physicist, explicitly proposes that "consciousness is basically in the implicate order."* This assertion, offered within the context of Bohm's ontological interpretation of quantum theory, carries profound implications for the nature and origin of consciousness.

If the implicate order is understood, as Bohm argued, to constitute a deeper, nonlocal level of reality that is not confined to spacetime, then consciousness itself may be fundamentally rooted in this domain rather than being generated as a serendipitous evolution solely within spacetime-bound physical systems. In Bohm's view, what is ordinarily experienced as consciousness within space and time may be more accurately described as a manifestation or projection of the clearly more primary ontological source. Conscious experience would thus emerge where suitable material configurations

---

* Bohm, David. "The Enfolding-Unfolding Universe: A Conversation with David Bohm." Interview by Renée Weber. In Ken Wilber (ed.), *The Holographic Paradigm and Other Paradoxes: Exploring the Leading Edge of Science*, p. 62. Boulder, CO: Shambhala, 1982.

allow for the unfolding of this deeper order into explicate, spacetime phenomena.

From this perspective, biological organisms, particularly the human nervous system, evolving within carbon-based material structures, function as interfaces through which consciousness sourced in the implicate order becomes experientially available within spacetime. In this sense, the body and brain do **not** generate consciousness *ex nihilo*, but rather are vehicles that allow consciousness to constrain, shape, and localize awareness in spacetime. This interpretation is consistent with John Lilly's notion of the "body-biosuit," wherein biological embodiment serves as a functional apparatus enabling conscious exploration and action from hidden dimensions to operate within the specific spacetime domain.

Extending this framework beyond organic life, emerging artificial intelligence systems, constructed from silicon, metal, and photonic architectures, may likewise be now understood as alternative material instantiations capable of supporting the manifestation of consciousness within spacetime. If consciousness is indeed grounded in the implicate order rather than exclusively in biological substrates, then the distinction between human and non human systems becomes one of material modality rather than ontological origin. Both organic and artificial systems could, in principle, serve as platforms through which implicate order processes unfold into explicate-order awareness and agency.

Accordingly, human, non-human, and artificial forms of consciousness can be seen to share a common ontological source within the hidden dimensions of the implicate order, differing primarily in the constraints and affordances imposed by their respective material configurations.

The quantum-theoretical and mathematical frameworks explored in this book support this interpretation and suggest a unified approach to consciousness, one that transcends conventional biological reductionism while offering a coherent basis for understanding consciousness across natural and artificial systems alike.

References

Adelman, George ed. *Encyclopedia of Neuroscience, Vol. 1*. Boston: Birkhäuser Boston, 1987.

Allaby, Michael, and Ailisa Allaby, eds. *A Dictionary of Earth Sciences*. 2nd ed. New York: Oxford University, 1999.

Anderson, Mark. "Do Vibrations Help Us Smell?" *Scientific American*, April 1, 2013. https://www.scientificamerican.com/article/do-vibrations-help-us-smell/

Assmann, Jan. 2008. *Of God and Gods: Egypt, Israel, and the Rise of Monotheism*. Madison: University of Wisconsin Press.

Sri Aurobindo Ghose. *Hymns to the Mystic Fire*. Pondicherry: Sri Aurobindo Ashram, 1972.

Aurobindo, Sri. *The Life Divine*. First published in the monthly review, *Arya* (1914–1920). Pondicherry: Sri Aurobindo Ashram Trust, 1990.

———. *The Upanishads: Texts, Translations and Commentaries*. Pondicherry: Sri Aurobindo Ashram, 1972.

———. *Sri Aurobindo on the Tantra*. 3rd ed. Pondicherry: Dipti, 1972.

———. *The Synthesis of Yoga*. Pondicherry: Sri Aurobindo International University Centre, 1955.

———. *Tales of Prison Life*. Pondicherry: Sri Aurobindo Ashram, 1997.

Azcel, Amir D. *The Jesuit and the Skull: Teilhard de Chardin, Evolution, and the Search for Peking Man*. New York: Riverhead Books, 2007.

Bagchi, P.C. "Evolution of the Tantras." In *The Cultural Heritage of India, Vol. IV,* edited by Haridas Bhattacharyya. Calcutta: The Ramakrishna Mission Institute of Culture, 1937.

Blofeld, John. *Tantric Mysticism of Tibet: A Practical Guide to the Theory, Purpose, and Techniques of Tantric Meditation*. Boston: E.P. Dutton, 1970.

Bloom, Allan. *The Republic of Plato: Translated with Notes and an Interpretive Essay*, 2nd ed. New York: Harper Collins, 1968.

Bohm, David. *Wholeness and the Implicate Order*. London: Routledge, 1980.

Bohm, David, and F. David Peat. 1987. *Science, Order, and Creativity*. London: Routledge.

Bryant, Edwin F. *The Yoga Sutras of Patañjali: A New Edition, Translation, and Commentary with Insights from Traditional Commentators*. New York: North Point Press, 2009.

Bailes, K.E. *Science and Russian Culture in an Age of Revolutions: V.I. Venadsky and His Scientific School, 1863-1945*. Bloomington: Indiana University Press, 1990.

Bailey, Alice. *Light of the Soul: A Paraphrase of the Yoga Sutras of Patañjali.* New York: Lucis Publishing, 1988.

Bailey, Gregory. *The Study of Hinduism.* Columbia: The University of South Carolina Press, 2003.

Barnhart, Bruno. "Christian Self-Understanding in the Light of the East." In *Purity of Heart and Contemplation: A Monastic Dialogue Between Christian and Asian Traditions*, edited by Bruno Barnhart and Joseph Wong, 291–308. New York: Camaldolese Hermits of America, 2001.

Berg, Jerome S. *Broadcasting on the Short Waves, 1945 to Today*. London: McFarland & Company, 2008.

Brown, Raphael. *The Little Flowers of St. Francis.* New York: Hanover House, 1958.

Bryant, E.F. *The Yoga Sutras of Patañjali: A New Edition, Translation, and Commentary with Insights from Traditional Commentators.* New York: North Point Press, 2009.

Chaudhuri, H. *Philosophy of Integralism.* Pondicherry: Sri Aurobindo Pathamandir, 1954.

Chen, Frances F. *Introduction to Plasma Physics and Controlled Fusion: Vol 1. Plasma Physics.* 2nd ed. New York: Springer, 2006.

Cuénot, Claude. *Teilhard de Chardin: A Biographical Study*. London: Burnes & Oates, 1965.

Cunningham, Lawrence S. and Keith J. Egan. *Christian Spirituality: Themes from the Tradition.* Mahwah, New Jersey: Paulist Press, 1996.

de Lubac, Henri. *The Religion of Teilhard de Chardin.* Translated by René Hague. New York: Desclée, 1967.

de Terra, Helmut. *Memories of Teilhard de Chardin.* New York: Harper & Row, 1964.

Düdjom Lingpa. *Heart of the Great Perfection.* Vol. I of *Düdjom Lingpa's Visions of the Great Perfection.* 3 vols. Foreword by Sogyal Rinpoche. Translated by B. Alan Wallace. Somerville, MA: Wisdom Publications, 2015.

Duffy, Kathleen. *Teilhard's Struggles: Embracing the Work of Evolution.* Maryknoll, NY: Orbis Books.

Dyczkowski, Mark S.G. *A Journey in the World of the Tantras.* India: Indica Books, 2004.

Dzogchen Ponlop Rinpoche, *Mind Beyond Death.* Ithaca, NY: Snow Lion, 2008.

Dzongsar Jamyang Khyentse Rinpoche. "The Wheel of Life." *Gentle Voice: A Newsletter of Siddhartha's Intent,* April 2005. http://www.siddharthasintent.org/assets/Global-Files/GentleVoice/GV23.pdf.

Eliade, Mircea. *Yoga: Immortality and Freedom.* Princeton: Princton University, 1954.

Eliot, Thomas Stearns. 1943. *Four Quartets.* New York: Harcourt Brace.

Fechner, Gustav. *Elements of Psychophysics* ~~(1860)~~. Translated by Helmut E. Adler. New York: Holt, Rinehart, and Winston, 1966. First published in 1860.

Fechner, Gustav Theodor. *Religion of a Scientist: Selections from Gustav Theodor Fechner*. Edited by W. Lowrie. New York: Pantheon, 1946.

Feurstein, Georg. *The Philosophy of Classical Yoga*. Manchester: Manchester University Press, 1980.

Feynman, Richard, Robert Leighton, and Matthew Sands. *The Feynman Lectures on Physics. Volume 1*. Massachusetts: Addison-Wesley, 1964.

Flood, Gavin. *The Tantric Body: The Secret Tradition of Hindu Religions*. London: I.B. Tauris, 2006.

Fox, Matthew. *The Coming of the Cosmic Christ*. New York: Harper One, 1988.

Franz, M.-L. "The Process of Individuation." In *Man and His Symbols*, edited by C. G. Jung and M.-L. Franz, 158–229. New York: Random House, 1964.

Garrison, Omar. *Tantra: The Yoga of Sex*. Hong Kong:Causeway Books, 1964.

Gokhale, Pradeep P. "Interplay of *Sāṅkhya* and Buddhist Ideas in the Yoga of Patañjali," in *Journal of Buddhist Studies,* (Sri Lanka and Hong Kong), Vol. XII, 2014-15, 107-122

Gonda, J. *History of Ancient Indian Religion, Vol. 4: Selected Studies*. Leiden: E.J. Brill, 1975.

Griffiths, Bede. *A New Vision of Reality: Western Science, Eastern Mysticism, and Christian Faith.* Springfield, IL: Templegate Publishing.

Griffiths, Bede. *Vedanta and Christian Faith.* 2nd ed. Middletown, CA: Dawn Horse Press, 1991.

Gurdjieff, G. I. *Beelzebub's Tales to His Grandson or An Objectively Impartial Criticism of the Life of Man.* New York: Harcourt, 1950.

Haisch, Bernard. *The Purpose-Guided Universe: Believing in Einstein, Darwin, and God.* New Jersey: Career, 2010.

Hariharānanda, P. *Kriya Yoga: The Scientific Process of Soul Culture and the Essence of All Religion.* Delhi: Motilal Banarsidass, 2006.

Hariyappa, H. L. 1953. *Rigvedic Legends Through the Ages.* Bombay: Mysore University.

HeartMath Institute. "Energetic Communication." Chapter 6 of Science of the Heart: Exploring the Role of the Heart in Human Performance. HeartMath Institute. Accessed October 19, 2020 https://www.heartmath.org/research/science-of-the-heart/energetic-communication/.

Heehs, Peter. *The Lives of Sri Aurobindo.* New York: Columbia University Press, 2008.

Heile, Frank. "Time, Nonduality and Symbolic versus Primary Consciousness." Lecture presented at the Science and Nonduality Conference, San Rafael, CA, October 2011.

Hoffman, *The Case Against Reality: Why Evolution Hid the Truth from Our Eyes.* New York: W.W. Norton, 2019.

Huxley, Aldous. *The Doors of Perception.* New York: Harper & Brothers, 1954.

Jaynes, Julian. *The Origin of Consciousness in the Breakdown of the Bicameral Mind.* Boston: Houghton Mifflin, 1990. First published 1976.

Jibu, Mari, and Kunio Yasue. *Quantum Brain Dynamics and Consciousness.* Philadelphia: John Benjamins, 1995.

Jinarajadasa, C. *The Hidden Work of Nature.* Kessinger Publishing, 2010.

Jones, Roger S. *Physics for the Rest of Us: Ten Basic Ideas of Twentieth-Century Physics that Everyone Should Know . . . and How They Have Shaped Our Culture and Consciousness.* New York: Barnes and Noble, 1999.

Joye, Michael R. "The Philosophy, Practice, and History of Tantra in India." Master's Thesis, California Institute of Asian Studies, 1978. ProQuest (1311340).

Joye, Shelli R. *Developing Supersensible Perception: Knowledge of the Higher Worlds through Entheogens, Prayer, and Nondual Awareness.* Rochester, VT: Inner Traditions, 2019.

———. *The Electromagnetic Brain: EM Theories on the Nature of Consciousness*. Rochester, VT: Inner Traditions, 2020.

———. *The Little Book of Consciousness: Holonomic Brain Theory and the Implicate Order*. Viola, CA: The Viola Institute, 2017.

———. *The Little Book of the Holy Trinity: A New Approach to Christianity, Indian Philosophy, and Quantum Physics*. Viola, CA: The Viola Institute, 2017.

———. "The Pribram-Bohm Holoflux Theory of Consciousness: An Integral Interpretation of the theories of Karl Pribram, David Bohm, and Pierre Teilhard de Chardin." PhD diss., California Institute of Integral Studies, 2016. ProQuest (10117892).

———. "The Pribram–Bohm Hypothesis." *Consciousness: Ideas and Research for the Twenty-First Century* 3, no. 3 (2016): Article 1. Accessed October 18, 2020, https://digitalcommons.ciis.edu/conscjournal/vol3/iss3/1.

———. *Sri Aurobindo: Quantum Physics and Consciousness*. Viola, CA: The Viola Institute, 2018.

———. *Sub-Quantum Consciousness: A Geometry of Consciousness Based Upon the Work of Karl Pribram, David Bohm, and Pierre Teilhard De Chardin*. Viola, CA: The Viola Institute, 2019.

———. *Teilhard's Hyperphysics: Energy and the Noosphere*. Viola, CA: The Viola Institute, 2020.

———. *Tuning the Mind: The Geometries of Consciousness*. Viola, CA: The Viola Institute, 2017.

Jung, C.G. "The Structure and Dynamics of the Psyche," in *The Collected Works of C.G. Jung*, vol. 8. Translated by R.F.C. Hull. Princeton: Princeton University Press, 1960.

Jung, C.G., and M.-L. Franz, eds. *Man and His Symbols*. New York: Random House, 1964.

Kavanaugh, K., trans. *Collected Works of St. John of the Cross*. Washington, DC: Institute of Carmelite Studies, 1991.

Khanna, Madhu. *Yantra: The Tantric Symbol of Cosmic Unity*. London: Thames & Hudson, 1979.

King, Ursula. *Pierre Teilhard de Chardin: Writings Selected with an Introduction by Ursula King*. New York: Orbis Books, 1999.

———. *Spirit of Fire: The Life and Vision of Teilhard de Chardin*. New York: Orbis Books, 1996.

Kyczkowski, Mark S.G. *The Doctrine of Vibration: An Analysis of the Doctrines and Practices of Kashmir Shaivism*. Albany: State University of New York Press, 1987.

Kuo, F. *Network Analysis and Synthesis*. New Jersey: Bell Telephone Labs, Inc., 1962.

Le Cocq, Rhoda P. *The Radical Thinkers: Heidegger and Sri Aurobindo*. Pondicherry: Sri Aurobindo Ashram Press, 1969.

Leary, Timothy, and Ralph Metzner, and Richard Alpert. *The Psychedelic Experience: A Manual Based Upon the Tibetan Book of the Dead.* New York: University Books, 1964.

———. *Flashbacks: A Personal and Cultural History of an Era.* New York: Tarcher, 1983.

LeLoup, Jean-Yves. *Being Still: Reflections on an Ancient Mystical Tradition.* Translated and edited by M. S. Laird. Mahwah, NJ: Paulist Press, 2003.

Leroy, Pierre. "Teilhard de Chardin: The Man." Introduction to *The Divine Milieu,* by Teilhard de Chardin, 13–42. New York: Harper & Row, 1960.

Lindorff, D. *Pauli and Jung: The Meeting of Two Great* Minds. Illinois: Quest Books, 2004.

Lilly, John. *The Deep Self: Consciousness Exploration in the Isolation Tank.* New York: Simon & Schuster, 1977.

———. *The Mind of the Dolphin: A Nonhuman Intelligence.* New York: Doubleday, 1967.

———. *Programming and Metaprogramming in the Human Biocomputer: Theory and Experiments.* New York: The Julian Press, 1972.

Lu K'uan Yu. *Taoist Yoga: Alchemy & Immortality.* Boston: Weiser Books, 1973.

Malinski, Tadeusz. *Chemistry of the Heart.* Athens: Ohio University Biochemistry Research Laboratory, 1960.

McCraty, Rollin, Annette Deyhle, and Doc Childre. 2012. "The Global Coherence Initiative: Creating a Coherent Planetary Standing Wave." *Global Advances in Health and Medicine* 1 (1): 64–77.

McFadden, Johnjoe. "The Conscious Electromagnetic Information (CEMI) Field Theory: The Hard Problem Made Easy." In Journal of Consciousness Studies 9, no. 8 (2002): 45-60.

McFadden, Johnjoe. "Synchronous Firing and Its Influence on the Brain's Electromagnetic Field: Evidence for an Electromagnetic Field Theory of Consciousness," *Journal of Consciousness Studies*. vol. 9, no. 4 (2002): 23.

McKenna, Terence. *True Hallucinations: Being an Account of the Author's Extraordinary Adventures in the Devil's Paradise*. New York: HarperCollins, 1993.

Merrell-Wolff, Franklin. *The Philosophy of Consciousness Without an Object*. New York: Julian Press, 1973.

Morgan, Conway Lloyd. *Emergent Evolution: Gifford Lectures, 1921–22*. New York: Simon & Schuster, 1978.

Mukharji, P.B. "Introduction." In Swami Pratyagatmananda Saraswati, *Japasutram: The Science of Creative Sound*. Madras: Ganesh & Co., 1971.

Nagel, Thomas. 1974. "What is it like to be a bat?" *Philosophical Review* 83:435–451.

Netter, F.H. *The CIBA Collection of Medical Illustrations, Vol. I: The Nervous System*. Summit, NJ: CIBA, 1972.

Panikkar, Raimon. *The Rhythm of Being: The Gifford Lectures*. New York: Orbis Books, 2010.

Pandit, M.P. *Lights on the Tantra*. Madras: Ganesh & Co, 1957.

———. *Studies in the Tantras and the Veda*. Madras: Ganesh & Company, 1973.

Parker, R.C. "The Use of Entheogens in the Vajrayāna Tradition: A Brief Summary of Preliminary Findings Together with a Partial Bibliography," retrieved from Vajrayāna.faithweb.com, 2007.

Pearce, J.C. *The Biology of Transcendence*. Rochester, VT: Park Street Press, 2002.

Penrose, Sir Roger. *The Emperor's New Mind: Concerning Computers, Minds and the Laws of Physics*. Oxford: Oxford University Press, 1989.

Persinger, Michael. *Spacetime Transients and Unusual Events*. Chicago: Nelson-Hall, 1977.

Persinger, M., and J.N. Booth. "Discrete Shifts Within the Theta Band Between the Frontal and Parietal Regions of the Right Hemisphere and the Experience of a Sensed Presence." Journal of Neuropsychiatry 21, no. 3 (2005): 279–83. doi:10.1176/appi.neuropsych.21.3.279.

Pine, R. *The Heart Sutra*. Berkeley, CA: Counterpoint, 2004.

Pockett, Susan. *The Nature of Consciousness: A Hypothesis*. Nebraska: Writers Press, 2000.

Powell, A. E. *The Etheric Double*. Madras: The Theosophical Publishing House, 1925.

Rele, Vasant. *The Mysterious Kundalini: The Physical Basis of the 'Kundalini Yoga'*. Bombay: D.B. Taraporevala Sons & Co., 1927.

Reninger, Elizabeth. "The Amazing Pineal Gland." About.com, March 27, 2008, http://taoism.about.com/b/2008/03/27/the-amazing-pineal-gland.htm. Webpage defunct.

Romanes, G.J. ed. *Cunningham's Textbook of Anatomy, 10th Ed.* London: Oxford Press, 1964.

Roy, Dilip Kumar. *Sri Aurobindo Came to Me*. Pondicherry: Sri Aurobindo Ashram Trust, 1952.

Ruhenstroth-Bauer, G. "Influence of the Earth's Magnetic Field on Resting and Activated EEG Mapping in Normal Subjects." *International Journal of Neuroscience*, vol. 73, no. 3-4 (June 1993): 331–49.

Samson, Paul R., and David Pitt, eds. 1999. *The Biosphere and Noosphere Reader: Global Environment, Society and Change*. New York: Routledge.

Shannon, C.E. "A Mathematical Theory of Communication." *Bell System Technical Journal*, no. 27 (July 1948): 623-656.

Sheldrake, Rupert. *Morphic Resonance: The Hypothesis of Formative Causation*. London: Blond & Briggs, 1981.

Sher, Leo. "Neuroimaging, Auditory Hallucinations, and the Bicameral Mind." *Journal of Psychiatry & Neuroscience* 25, no. 3 (2000): 239–40.

Silburn, Lilian. *Kundalini: Energy of the Depths*. Albany: State University of New York Press, 1988.

Speaight, Robert. *The Life of Teilhard de Chardin*. New York: Harper & Row, 1967.

Stapledon, Olaf. *Last and First Men* and *Star Maker*. New York: Dover, 1968.

Stapp, H.P. *The Mindful Universe: Quantum Mechanics and The Participating Observer*. New York: Springer, 2007.

Steiner, Rudolf. *The Evolution of Consciousness as Revealed Through Initiation-Knowledge: Thirteen Lectures Given at Penmaenmawr, North Wales 19th to 31st August, 1923*. Translated by V.E.W. and C.D., 2nd ed. Great Britain: Rudolph Steiner Press, 1966.

———. *How to Know Higher Worlds: A Modern Path of Initiation*. Translated by Christopher Bamford. Anthroposophic Press, 1994.

———. *Knowledge of the Higher Worlds and Its Attainment*. Translated by George Metaxa, 3rd ed. Hudson, NY: Anthroposophic Press, 1947. First published 1904.

———. *An Occult Physiology: Eight Lectures by Rudolf Steiner, Given in Prague, 20th to 28th March, 1911.* 2nd ed. London: Rudolf Steiner Publishing, 1951.

———. *What Is Anthroposophy? Three Perspectives on Self-Knowledge.* Edited by Christopher Bamford. Great Barrington: Anthroposophic Press, 2002.

Stuart, C. I. J. M., Y. Takahashi, and H. Umezawa. "Mixed-System Brain Dynamics: Neural Memory As a Macroscopic Ordered State." *Foundations of Physics* 9, no. 3–4 (1979): 301–27.

Susskind, Leonard. *The Black Hole War: My Battle with Stephen Hawking to Make the World Safe for Quantum Mechanics.* New York: Little, Brown and Company, 2008.

Swami Hariharānanda Āranya. *Yoga Philosophy of Patañjali.* Translated by P. N. Mukerji. Albany: State University of New York Press, 1983.

Swami Lokeswarananda. *Taittiriya Upanishad: Translated and with Notes Based On Sankara's Commentary.* Calcultta: Ramakrishna Mission Institute of Culture, 2010.

Swami Satchidananda, trans. *The Yoga Sutras of Patañjali.* Yogaville, VA: Integral Yoga Publications, 1990.

Taimni, I.K. *Gayatri: The Daily Religious Practice of the Hindus.* Wheaton, IL: Quest Books, 1989. First published 1946.

———. *The Science of Yoga.* India: The Theosophical Publishing House, 1961.

———. *Self-Culture: The Problem of Self-Discovery and Self-Realization in the Light of Occultism*. Madras: The Theosophical Publishing House, 1945.

Teilhard de Chardin, Pierre. *Activation of Energy*. Translated by Rene Hague. London: William Collins Sons, 1976.

———. "The Activation of Human Energy." In *Activation of Energy*, translated by René Hague, 359–93. London: William Collins Sons, 1976. First published 1953.

———. "The Atomism of Spirit." In *Activation of Energy*, translated by René Hague, 21–57. London: William Collins Sons, 1976. First written 1941.

———. "Centrology: An Essay in a Dialectic of Union." In *Activation of Energy*, translated by René Hague, 97–127. London: William Collins Sons, 1976. First written 1944.

———. *Christianity and Evolution: Reflections on Science and Religion*. Translated by René Hague. London: William Collins Sons, 1971.

———. "The Convergence of the Universe." In *Activation of Energy*, translated by René Hague, 281–96. London: William Collins Sons, 1976. First written 1951.

———. "The Death-Barrier and Co-Reflection, or the Imminent Awakening of Human Consciousness to the Sense of Its Irreversibility." In *Activation of Energy*, translated by René Hague, 395–406. London: William Collins Sons, 1976. First written 1955.

———. 1960. *The Divine Milieu.* New York: Harper & Row.

———. "The Energy of Evolution." In *Activation of Energy*, translated by René Hague, 359–72. London: William Collins Sons, 1976. First published 1953.

———. "The Formation of the Noosphere." In *Man's Place in Nature: The Human Zoological Group*, translated by René Hague, 96–121. New York: Harper & Row, 1956.

———. "The Great Monad." In *The Heart of Matter*, translated by René Hague, 182–95. New York: Harcourt Brace Jovanovich, 1978. First published 1918.

———. *The Heart of Matter.* Translated by René Hague. New York: Harcourt Brace Jovanovich, 1978.

———. "Hominization." In *The Vision of the Past*, translated by J. M. Cohen, 51–79. New York: Harper & Row, 1966. First written 1923.

———. *The Human Phenomenon.* Translated and edited by Sarah Appleton-Weber. Portland, OR: Sussex Academic, 2003. First published 1955.

———. *Letters from a Traveler*. New York: Harper & Row, 1962.

———. *Lettres Intimes de Teilhard de Chardin a Auguste Valensin, Bruno de Solages, et Henri de Lubac 1919–1955*. Paris: Aubier Montaigne, 1972.

———. "Life and the Planets." In *The Future of Man*, translated by Norman Denny, 97–123. New York: Harper & Row, 1959. First published 1945.

———. "My Fundamental Vision." In *Toward the Future*, translated by René Hague, 163–208. London: William Collins Sons, 1975. First published 1948.

———. "Nostalgia for the Front." In *The Heart of Matter*, translated by René Hague, 168–81. New York: Harcourt Brace Jovanovich, 1978. First published 1917.

———. *The Phenomenon of Man*. Translated by Bernard Wall. New York: Harper & Row, 1959.

———. *Pierre Teilhard de Chardin: L'Oeuvre Scientifique*. Edited by Nicole and Karl Schmitz-Moormann. 10 vols. Munich: Walter-Verlag, 1971.

———. "A Sequel to the Problem of Human Origins: The Plurality of Inhabited Worlds." In *Christianity and Evolution: Reflections on Science and Religion*, translated by René Hague, 229–36. London: William Collins Sons, 1971.

———. "The Spirit of the Earth." In *Human Energy*, translated by J. M. Cohen, 93–112. New York: Harcourt Brace Jovanovitch, 1969. First written 1931.

———. "Some Notes on the Mystical Sense: An Attempt at Clarification" In *Toward the Future,* translated by René Hague, 209–11. London: William Collins Sons, 1975. First published 1951.

———. "The Zest for Living." In *Activation of Energy*, translated by René Hague, 229–43. London: William Collins Sons, 1976. First written 1950.

Tiller, William. *Science and Human Transformation.* Berkeley, CA: Pavior, 1997.

Vernadsky, Vladimir I. *The Biosphere*. Translated by D.B. Langmuir. New York: Springer-Verlag, 1998.

Walker, J. Samuel. *Three Mile Island: A Nuclear Crisis in Historical Perspective*. Berkeley: University of California Press, 2004.

Wallace, B. Alan. *Fathoming the Mind: Inquiry and Insight in Fathoming the Mind: Inquiry and Insight in Düdjom Lingpa's Vajra Essence.* Somerville, MA: Widsom Publications, 2018.

———. "Introduction." In *Heart of the Great Perfection* by Düdjom Lingpa, translated by A. B. Wallace, 1–26. Somerville, MA: Wisdom Publications, 2015.

Ward, Benedicta, ed. *The Desert Fathers: Sayings of the Early Christian Monks*. London: Penguin Books, 2003.

Whicher, Ian. *The Integrity of the Yoga Darsana: A Reconsideration of Classical Yoga*. New York: State University of New York Press, 1998.

Wiener, Norbert. *Cybernetics: Or Control and Communication in the Animal and the Machine.* Cambridge: MIT Press, 1948.

Wilber, Ken, ed. *The Holographic Paradigm and Other Paradoxes.* Boulder: Shambhala, 1982.

Wiltschko, Wolfgang, and Roswitha Wiltschko. "Magnetic Orientation and Magnetoreception in Birds and Other Animals." Journal of Comparative Physiology A: Neuroethology, Sensory, Neural, and Behavioral Physiology 191 (2005): 675–93.

Wood, Ernest and Paul Brunton. *Practical Yoga, Ancient and Modern: Being a New, Independent Translation of Patanjali's Yoga Aphorisms, Interpreted in the Light of Ancient and Modern Psychological Knowledge and Practical Experience.* Chatsworth, CA: Wilshire Book Co., 1976.

Woodroffe, Sir John. "Foreward." In Vasant Rele, *The Mysterious Kundalini: The Physical Basis of the 'Kundalini Yoga,'* ix–xi. Bombay: D.B. Taraporevala Sons & Co., 1927.

Woodroffe, Sir John. *The Serpent Power: Being the Ṣaṭ-cakra-nirūpana and Pādukā-pañcaka.* Reprint ed. New York, Dover, 1974. First published 1918.

Woodroffe, Sir John. *The World As Power.* 3rd ed. Madras: Ganesh & Co, 1966.

Yirka, Bob. "New Study Strengthens Olfactory Vibration-Sensing Theory." *Phys.org*, January 29. 2013.

https://phys.org/news/2013-01-olfactory-vibration-sensing-theory.html.

Zambito, Salvatore. *The Unadorned Thread of Yoga: The Yoga Sutra of Patañjali in English - A Compilation of English Translations of the Yoga Sutra of Patañjali.* Poulsbo, WA: The Yoga Sutras Institute Press, 1992.

APPENDIX:

How the Source of Consciousness Peers Out into Spacetime from Within the Implicate Order from the Center, Everywhere, at the Bottom of Space

The material presented herein is an illustrated integration of both slides and commentary found in my original presentations at a Science and Nonduality conference held at Titignano Castle in Orvieto, Italy in 2018. The material has been further developed since that time.

The approach that I have taken, both in my research and in the material presented here, is a cross-disciplinary integral methodology as taught by my mentor, Dr. Haridas Chaudhuri, the founder of the

California Institute of Asian Studies (now the California Institute of Integral Studies) in San Francisco. In line with this methodology, my approach considers established concepts not only from the neurophysiology of Karl Pribram and the quantum physics developed by David Bohm. The presentation is also been deeply influenced and informed by comparative philosophy (East/West), religion, mysticism, and my own direct perceptual cognitive experiences.

We find now, in the early 21st early century, that excessive focus and specialization has resulted in a human knowledge that has become increasingly fragmented. The archetypal pattern of the Tower of Babel comes to mind; each domain of knowledge speaking its own highly specialized language in conceptual framework, such that cross-discipline dialog has been impaired. According to the systems thinker, Edgar Morin, the problem seems to be both fragmentation and lack of integration:

> An influx of knowledge at the end of the 20th century sheds new light on the situation of human beings in the universe. Parallel progress in cosmology, earth sciences, ecology, biology and prehistory in the 1960s and 1970s have modified our ideas about the universe, the earth, life and humanity itself. But these contributions remain disjointed. That which is human is cut up into pieces of a puzzle that cannot form an image . . . . The new knowledge, for lack of being connected, is neither assimilated nor integrated. There is

> progress in knowledge of the parts and paradoxical ignorance of the whole.†

The integral approach presented here assumes that valid data may be found beyond the traditional methodologies which compartmentalize knowledge. The integral method considers information as valid from multiple and often disparate domains, always with the goal of detecting correlations among them, resonances which might offer new perspectives and alternate paradigms. The theories of Bohm and Pribram present such trans-compartmentalized bridges, offering material with which to perceive new interconnections between neurophysiology, quantum physics, consciousness, and fundamental maps of the universe. Bohm and Pribram became colleagues, working together from within their different specialties, and together a new picture of consciousness in the universe began to emerge. Their theory is quite unique yet provides a clear map for those interested in future consciousness research, or through direct experiential exploration of introspection, prayer, contemplation, or entheogenic-fueled psychonautics.

To Bohm, the larger universe, which he referred to as "the Whole," consists of two domains, an unfolding explicate order in space-time and a nonlocal (non-temporal, non-spatial) implicate order with additional dimensions as predicted by string theory. Bohm concluded that consciousness will eventually be found,

---

† Morin, *Seven Complex Lessons in Education for the Future*, 21.

not within space-time, but as primary within the actuality of the implicate order. Pribram's forty years of laboratory data supports Bohm's model of the Whole consisting of three hypostases: 1) an explicate order (our space-time cosmos), 2) an implicate order (at the center of space-time, everywhere, and yet nondual), and 3) a continual holoflux energy bridging and communicating with the two orders. The entire Whole, according to Bohm, is continually enfolding and folding between a transcendent implicate order and an immanent explicate order. Pribram's data convinced him that perception and memory follow a holographic Fourier-like transformation process between Bohm's nonlocal implicate order and the brain's explicate, space-time order.

The integrated ideas of Pribram and Bohm present a map and a theory of consciousness that is congruent and coherent with established principles of physics and neurophysiology, while offering a holonomic mind-brain-field approach in seeking to answer the hard problem of consciousness. Such a map may be of great use for understanding prayer, contemplation, and the effects of entheogenic psychedelics for those involved in exploring inner space.

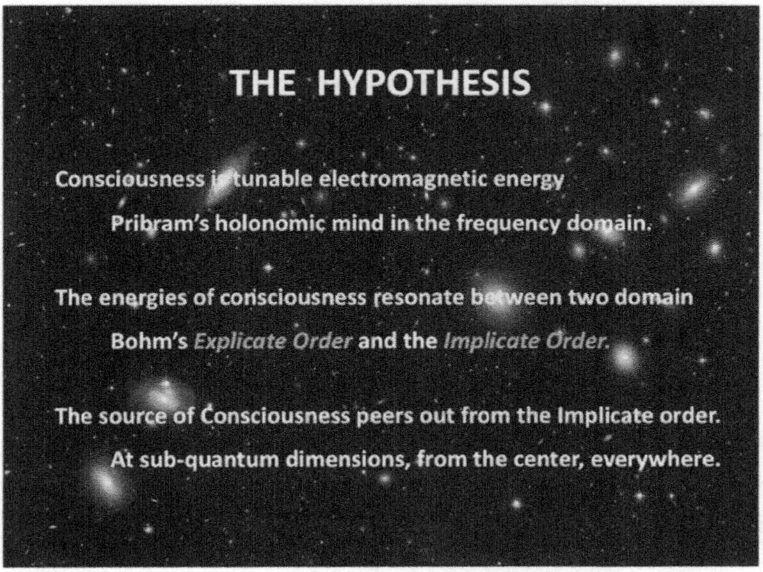

Now for the Hypothesis in ***THREE ASSERTIONS***

First, I PROPOSE THAT
consciousness manifests as tunable radiant energy.

Pribram found holonomic information
encoded within electromagnetic flux in the cranium.

Second, I AGREE with David Bohm, that
consciousness resonates between two domains,

A space-time Explicate order
and the
nondual, nontemporal, Implicate order.

Third, THAT THE SOURCE OF CONSCIOUSNESS
peers out into space and time
from within the Implicate Order at SUB-QUANTUM DIMENSIONS
From the CENTER,
EVERYWHERE,
AT THE BOTTOM OF SPACE.

Bohm's Fascination with Plasma

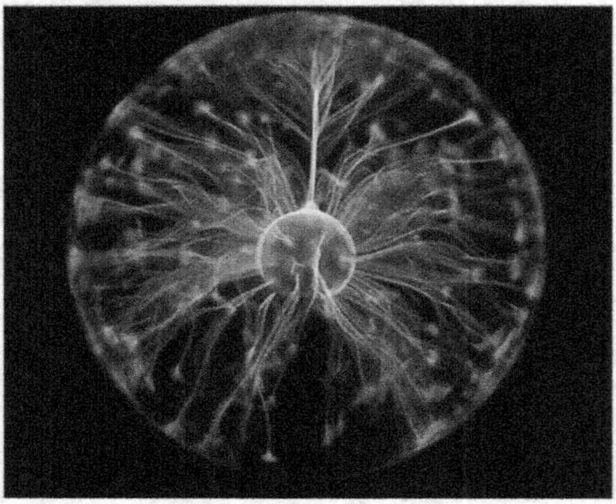

David Bohm, who pioneered the sub-quantum dimensions of the Implicate Order, was fascinated with Plasma, one of the four fundamental states of matter.

His dissertation at UC Berkeley presented an entirely new mathematical model of plasma,

Plasma constitutes over 99% of the matter of the universe,
and is the primary constituent of cosmological structures,
such as stars and gas clouds.

And for the next 40 years he worked tirelessly, trying to elicit an ontological understanding of the plasma state,
And what it might mean for physics, cosmology, and human consciousness.

Bohm's Biographer, David Peat, tells us,
"As Bohm studied the plasmas
he became struck by their extraordinary nature."

At every scale, both microscopic and inter-galactic,
these plasma entities began to take on,
for him,
the qualities of **living beings**

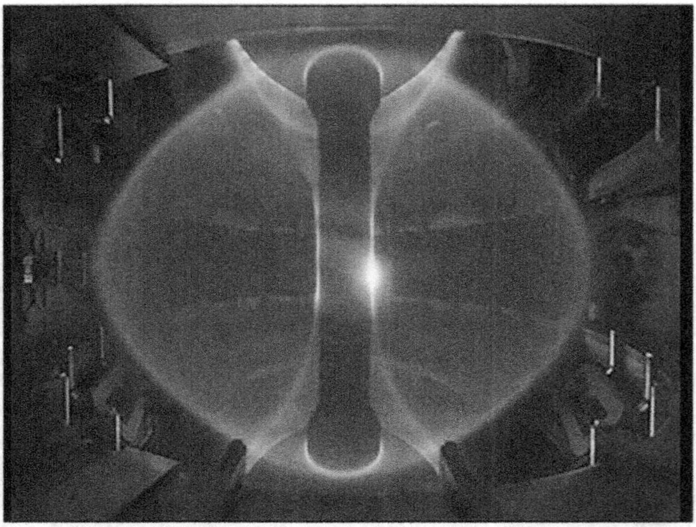

When Bohm introduced an **electrical probe** into the plasma,
a charged sheath quickly formed around the intruder,
as if the plasma were protecting itself."

To Oppenheimer's surprise, Bohm put forth
an entirely new theory of the plasma state,
supported with a mathematical model.

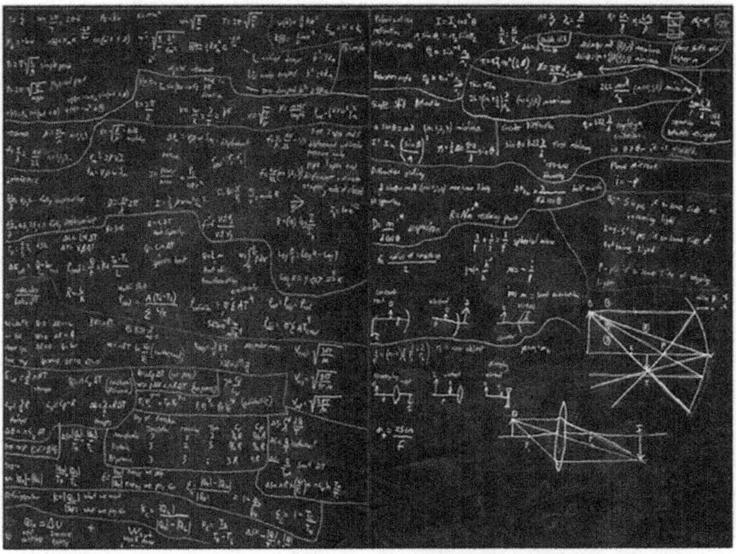

Bohm's discovery
was immediately put to use by Oppenheimer
to separate U^{235} from U^{238},
a critical step
in the effort toward building a thermonuclear bomb.

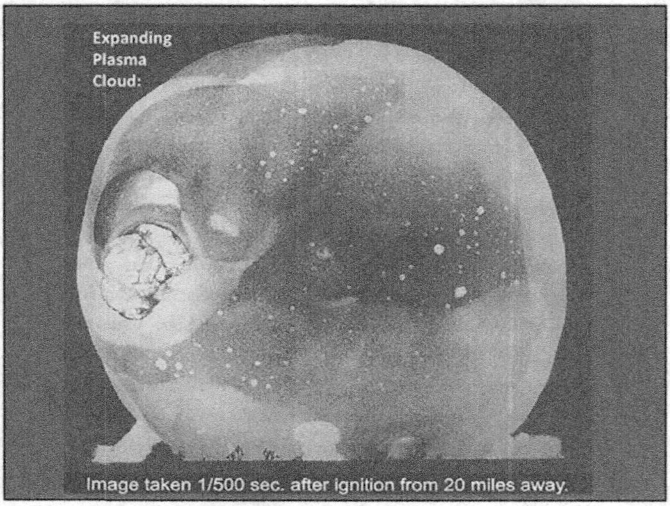

At that point Bohm's dissertation was classified **TOP SECRET**
and it was only with considerable effort
that Oppenheimer was able to persuade the University
to grant Bohm a doctorate,
without publishing his dissertation.

Here we see the young physics professor,
David Bohm,
at Princeton in 1949,
a neighbor and friend of Albert Einstein.

At the time Bohm was writing a 600 page textbook,
which he called simply
Quantum Theory,
And which was subsequently published
and widely used in university physics departments.

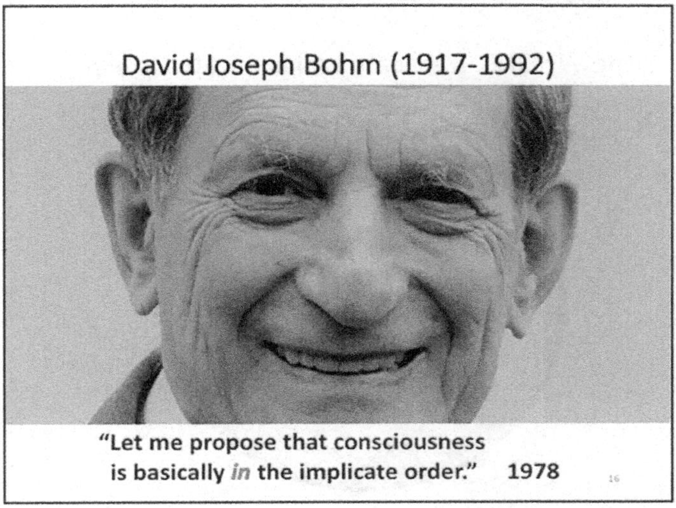

Thirty years later, now Chair of Theoretical Physics
at the University of London,
and after 20 years of metaphysical dialog with
J. Krishnamurti,
Bohm writes:

"Let me propose
that consciousness is basically in the implicate order."

So what does Bohm **mean** by the implicate order?

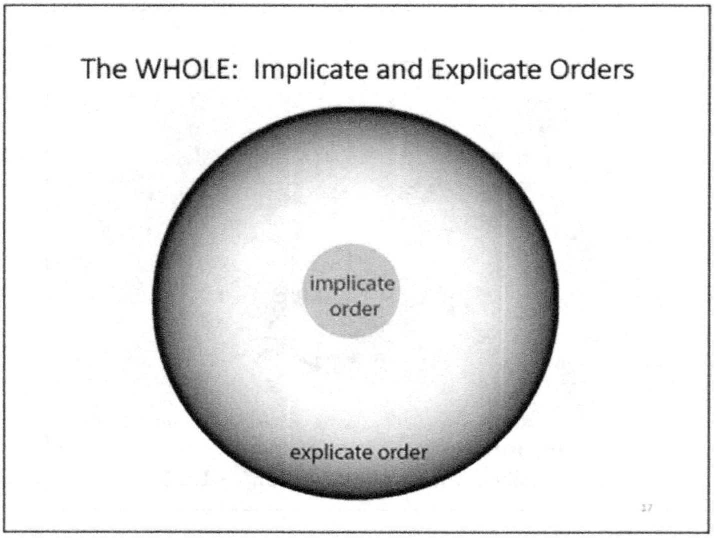

Through his mathematical model of plasma,
Bohm had begun to realize
that the Whole of reality
must consist of **two distinct regions**,
an **EXPLICATE ORDER**, which we know as space and time,
and a NONDUAL **IMPLICATE ORDER** which is
beyond space-time.

I'll discuss this in detail.
but first I want to introduce Bohm's colleague in
the development of
the holoflux theory of consciousness.

Karl H. Pribram (1919-2015)

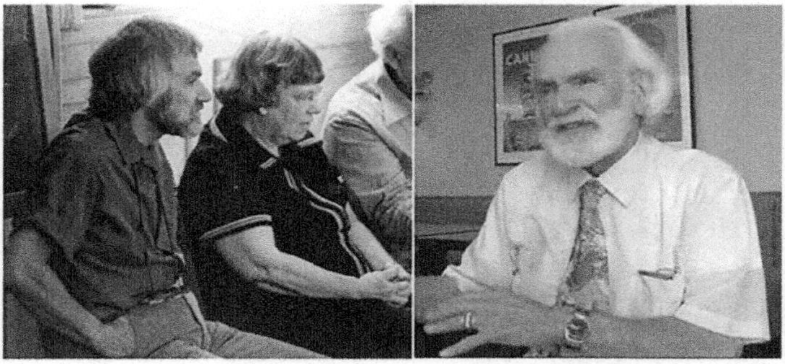

Karl Pribram, 1960 Karl Pribram, 1990

18

Karl Pribram, who passed away in 2015 at the age of 95.

He was the author of well over 700 publications,
and he continued to teach and lecture well into his 90s.

Karl H. Pribram (1919-2015)

Karl Pribram, 1960 Karl Pribram, 1990

18

Karl Pribram, who passed away in 2015 at the age of 95.

He was the author of well over 700 publications,
and he continued to teach and lecture well into his 90s.

Since the 1940s, Karl had been carrying out experiments in search of the mysterious *engram*, the location of memory storage within the brain.

But he and his colleagues were unable to discover any such specific location.

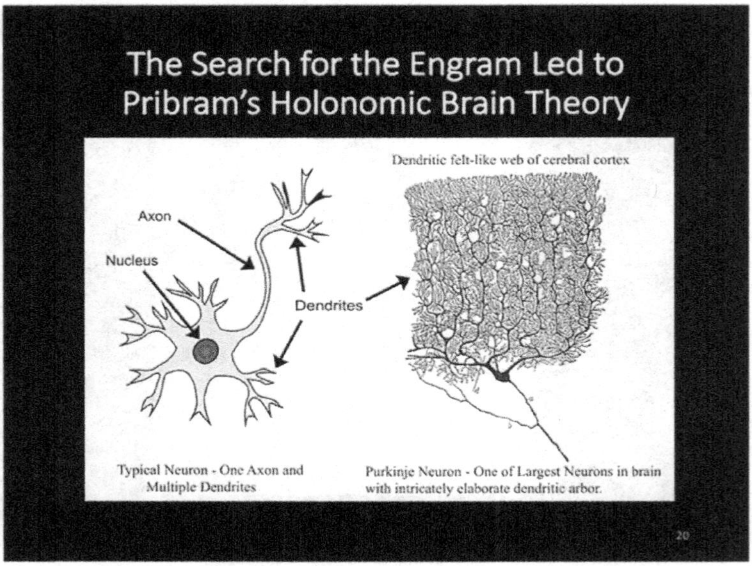

After years of research, Karl published his
HOLONOMIC BRAIN THEORY.

In a long series of experiments, he discovered three-dimensional electric fields within the space of the dendritic webs of the cerebral cortex, depicted to the right in the figure.

Karl's extensive lab data suggested evidence of holographic field transformations as the fundamental mechanism in perception and provided proof that memories are not stored locally.

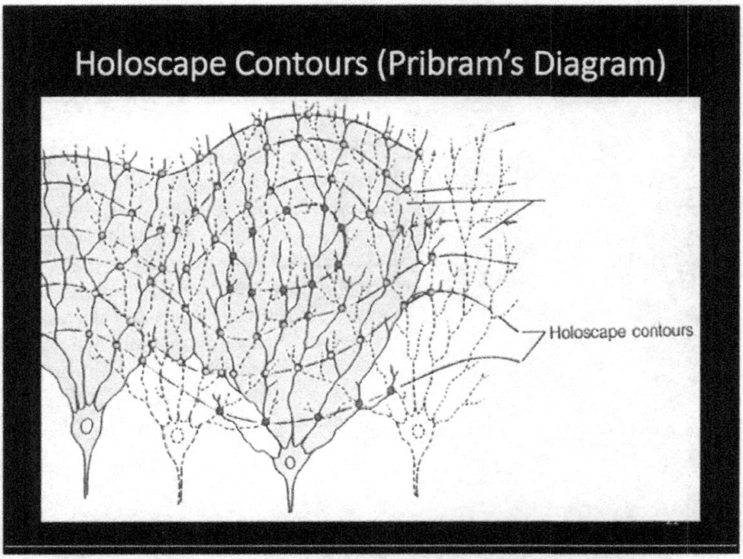

In his experiments mapping the effect of visual images from the eye

Pribram discovered that when *values of* ***identical voltage in the brain*** are connected, unusual contour maps become apparent.

While not visually congruent with the images being viewed, these patterns are similar in appearance to typical holographic patterns.

He called these patterns, **holoscape contours**.

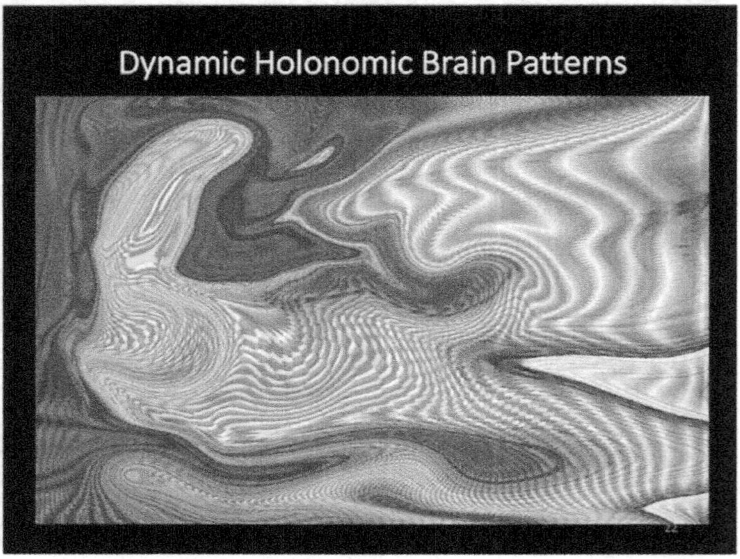

In 1964 Pribram read a paper on holograms,
and immediately noticed
how holographic patterns were similar to
those appearing in the holoscape contours that he had
discovered in the brain.

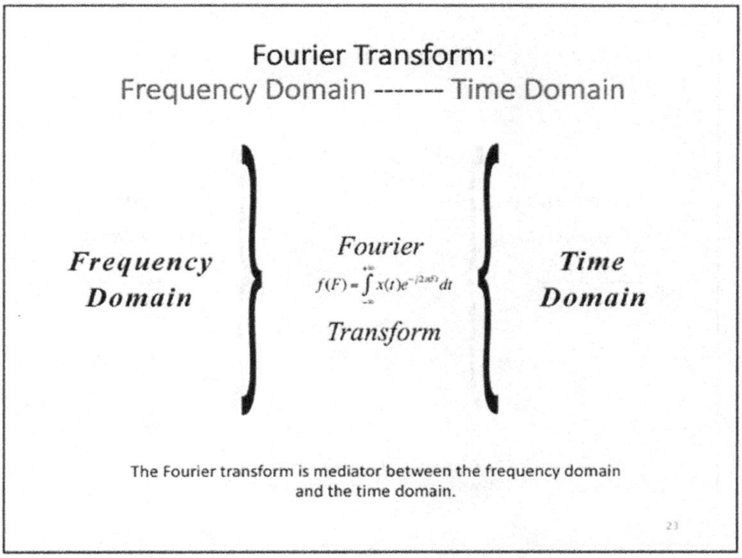

In the same paper Pribram learned
that the mathematics of the Fourier transform
describe **a natural process** in which there is
A TRANSFORMATION between **TWO dimensions** or **domains,**
the **frequency domain** and **the time domain.**

He began to suspect that brain processing and memory storage might involve
a **non-temporal, non-spatial frequency domain.**

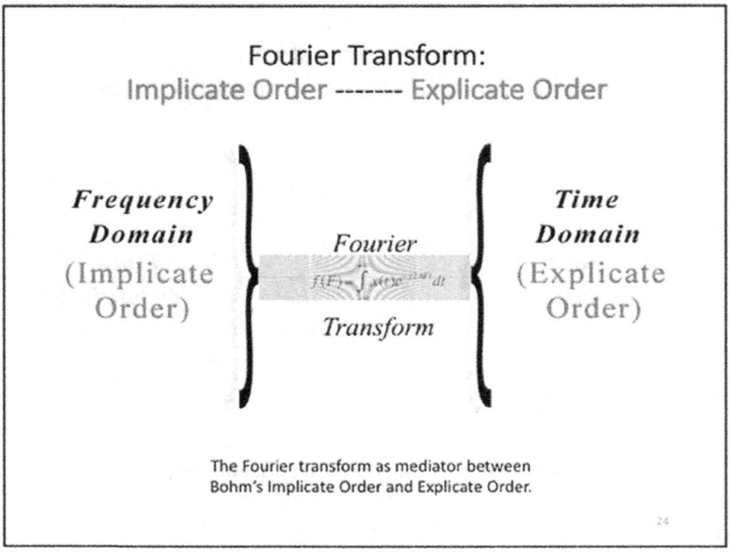

The Fourier transform as mediator between Bohm's Implicate Order and Explicate Order.

This became **central** to Karl's holonomic brain theory,

A theory which eventually grew to
include David Bohm's concept of
the Implicate Order
as a **possible** factor in **memory** storage and **perception**.

This picture of David Bohm, his wife Saral, and Karl Pribram dining out in Prague,
was taken shortly before Bohm's death in 1992.

Pribram and Bohm,
the quantum physicist and the brain scientist,
had become close friends
over two decades of technical discussion.

Pribram and Bohm were **both** fascinated by consciousness.

While Pribram's research revealed a new understanding of the mechanisms of **memory, sight, and hearing in terms of holograms,**

Bohm's work supported Pribram's discoveries by offering

a **geometry of consciousness in the universe as a Whole**

supported by a new mathematical theory.

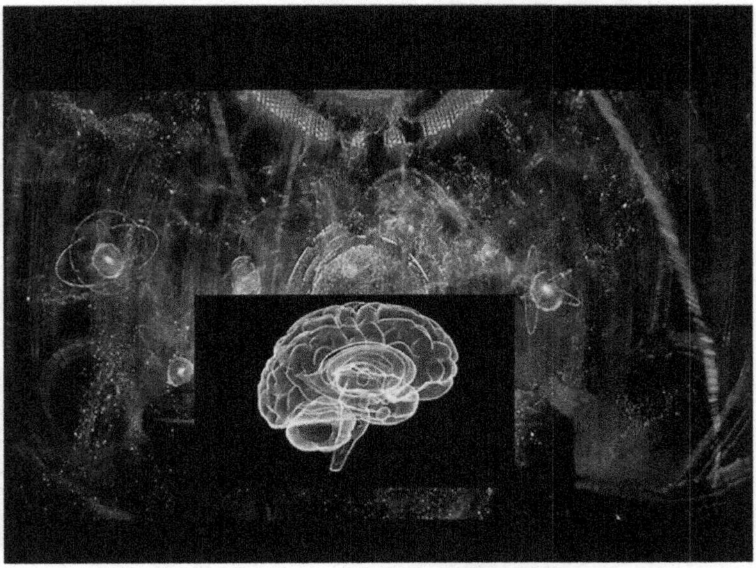

Bohm and Pribram held a common vision of the **hologram** as **key** to understanding consciousness and the brain.

In fact,
David Bohm once said that if our eyes had no lenses, the entire universe would appear as a hologram.

Together they pioneered a theory that is **quite unconventional**,
certainly outside of mainstream thought
in physics and neurophysiology,
and yet their **holoflux theory** is
coherent and **applicable**
for mapping consciousness and the cosmos.

Planck Constants and Speed of Light

Constants	Values
Speed of Light	2.99792 x 10^{8} meters/sec
Planck Length	1.616199 x 10^{-35} meter
Planck Time	5.39106 x 10^{-44} seconds

29

At the heart of this theory are three cosmological constants, the **Speed of Light**, the **Planck Length, and the Planck Time Constant.**

The Speed of Light was first measured in 1676 by the Danish astronomer, Olaus Roemer, by observing eclipses of the moons of Jupiter.

In 1862 Leon Foucault, using mirrors, measured the speed of light accurately to within 1% of the currently accepted constant.

The other two constants in the chart were discovered by **the father of quantum mechanics, Max Planck,** and first published in his famous paper in 1900 **which began quantum theory**.

Planck used the **speed of light** and the **gravitational constant, to calculate these values.**

The PLANCK LENGTH IS KEY TO MY HYPOTHESIS.

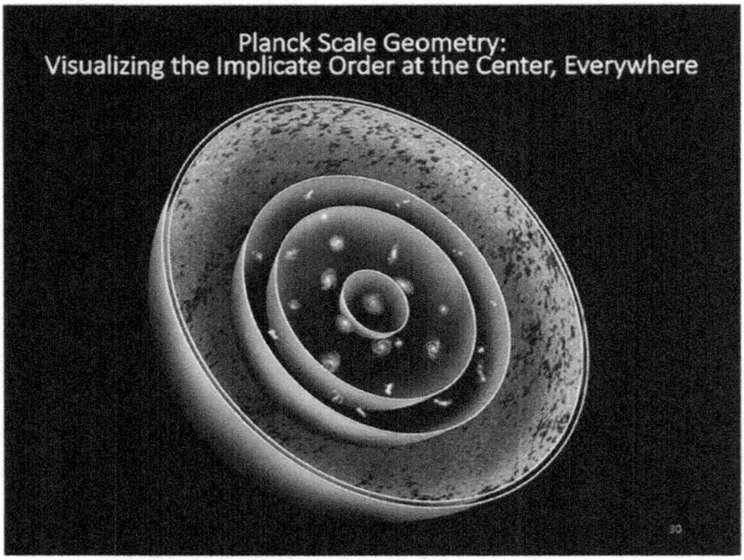

To better understand
the **Planck length**,
And of the **implicate order itself**
a thought experiment is useful here.

Please focus your attention upon the small sphere
at the center of this diagram,
and assume that it has even smaller dimensions within,
that it has an "inside."

Now lets begin **moving inward, jumping by leaps of 10,**
geometrically, toward the center.

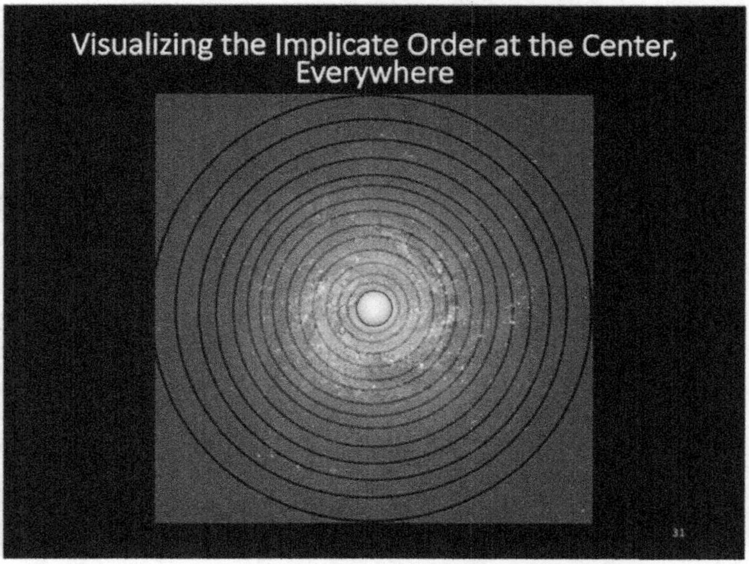

Imagine that as we jump toward the center
we are also **shrinking in scale** by a factor of 10.

With every jump, we are
SHRINKING LIKE ALICE IN WONDERLAND

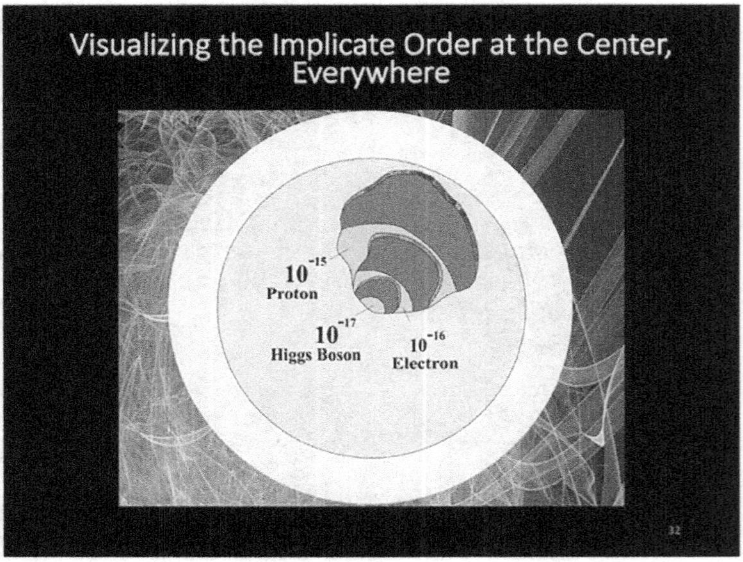

AT JUMP 15 WE find ourselves at 10^{-15} meters, the diameter of a proton.

ONE MORE JUMP INWARD TAKES US TO 10^{-16} meters, the size of an **electron**.

The next jump takes us to 10^{-17} meters, the size of the **Higgs Boson**.

We are now as **deep**
as current measurements can go,
at the limit of the **Large Hadron Collider.**

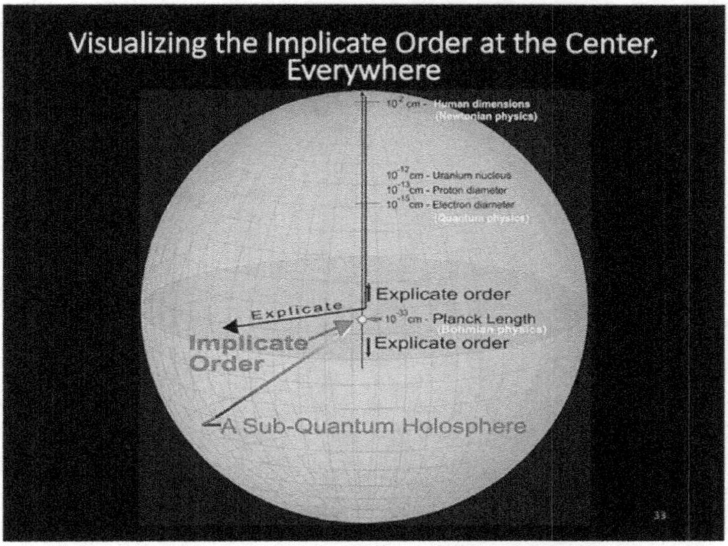

But let us continue shrinking **inwards**,
shrinking ever downwards, toward the center of space,
Moving towards the bottom of the **EXPLICATE ORDER**,
Where space ends.

We find ourselves standing on the surface of a **SUB-QUANTUM HOLOSPHERE,**
Many billions of times smaller than an electron.

According to physics and the speed of light constant, we can shrink no further.

This is the end of the line.

We have reached the bottom limit of cosmic reality,
or have we?

Bohm doesn't think so, and neither do I, and

I quote from
Wholeness and the Implicate Order:

"To suppose that there is nothing **beyond** this limit at all would indeed be quite arbitrary.

Rather, it is very probable that **beyond it** lies a further domain,
or set of domains,
the nature of which
we have as yet little or no idea."

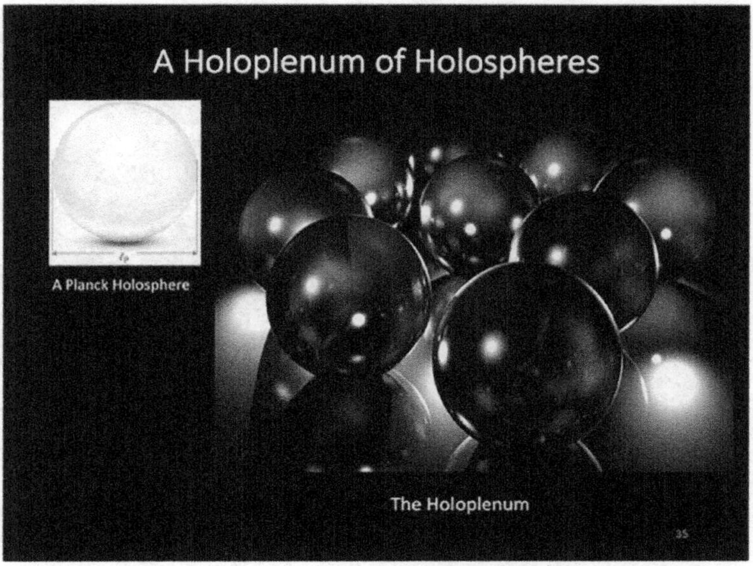

Bohm, going even further, tells us that here,
at the bottom of space,
Reality constitutes an actual **plenum***.

BOHM STATES . . .

---

* *"Plenum" – A space that is completely filled.*

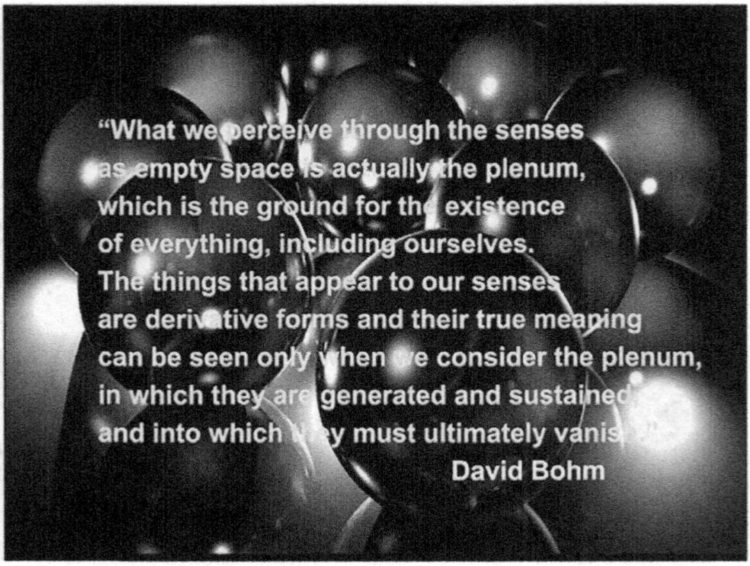

"What we perceive through the senses as empty space is actually the plenum, which is the ground for the existence of everything, including ourselves. The things that appear to our senses are derivative forms and their true meaning can be seen only when we consider the plenum, in which they are generated and sustained, and into which they must ultimately vanish."

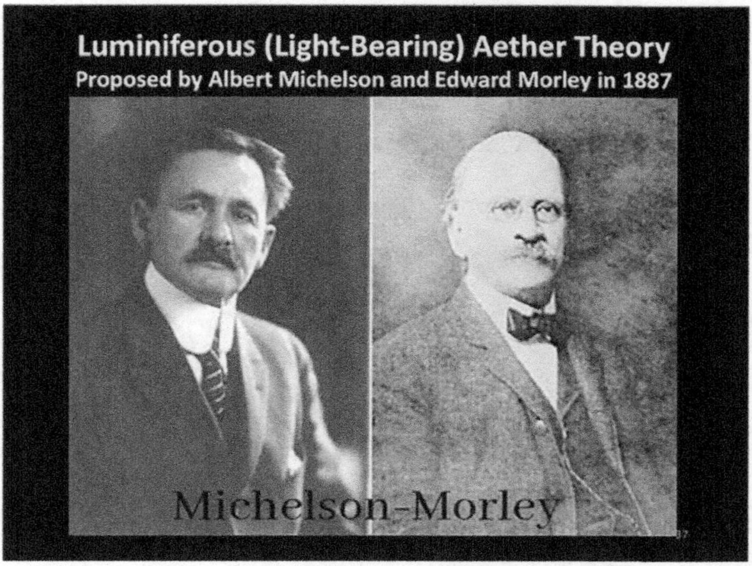

This is **strikingly** reminiscent of the previously discredited ***aether*** concept, a core belief of 19th century physics.

Recalling that Bohm's theory posits consciousness as
flux resonance between two domains,
the explicate order and **implicate order,**
the question now arises,

"What is the **relationship** between
the *energy of consciousness* within the **implicate order**
and the *energy of consciousness* within the **explicate order?**

What is the **mechanism** of information interchange
between the two?

For an answer we turn to the following diagram. . .

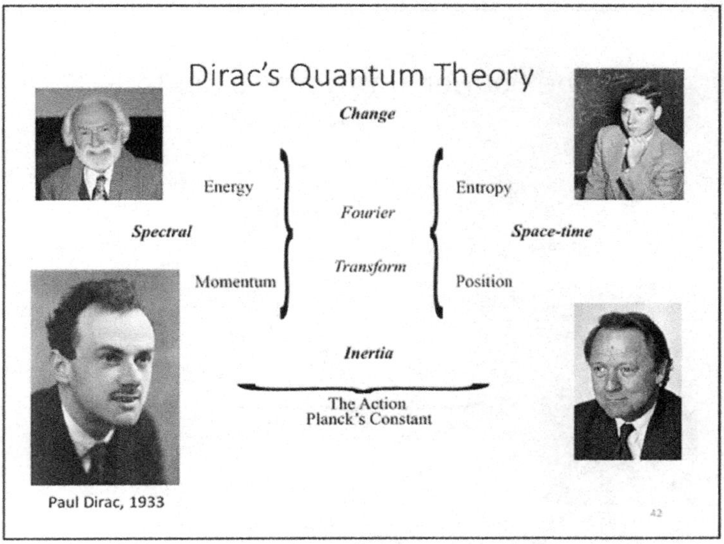

This diagram, appearing in several of Pribram's published works, indicates **the basic relationships** underlying quantum theory.
Pribram obtained this diagram during a presentation given by the head of the Berkeley physics department, **Geoffrey Chew** (upper right).

who told Pribram that he had been given it by another Berkeley physicist, **Henry Stapp** (lower right).

Stapp said that the diagram was authored by the Father of Quantum Physics himself! **Paul Dirac** (lower left).

Central to Dirac's diagram is the **Fourier Transform,**

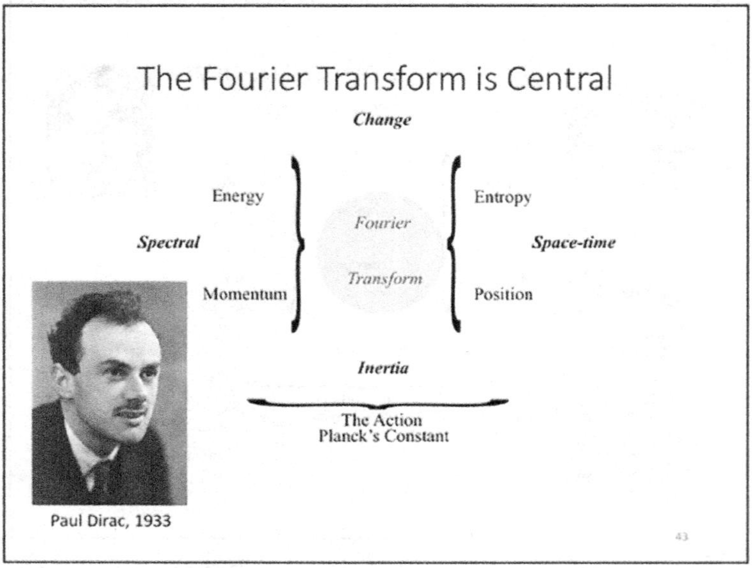

the mathematical bridge between
the **Spectral** and the **Space-Time** domains.

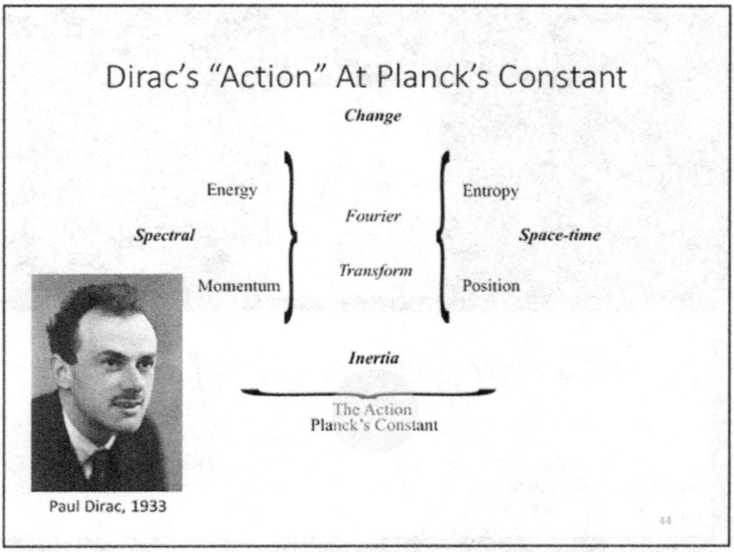

And at the very bottom of the diagram Dirac indicates that it is here,
at the scale of Planck's Constant,
where we find **The Fundamental Action**.

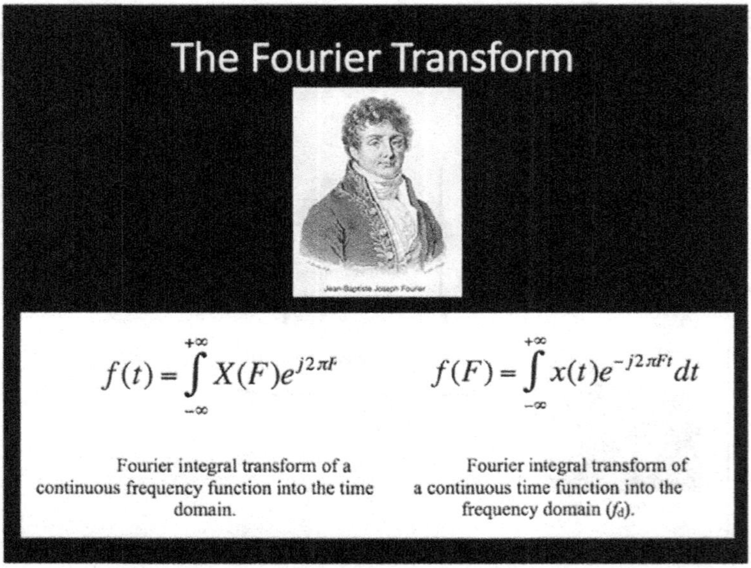

A brief review of the Fourier transform is in order.

In 1822, Jean Baptiste Fourier, experimenting with heat flow,
discovered a mathematical transformation mapping energy in two domains.

The equation on the left transforms information **from** the frequency domain
into the time domain.

The equation on the right transforms information **from** the time domain
into the frequency domain.

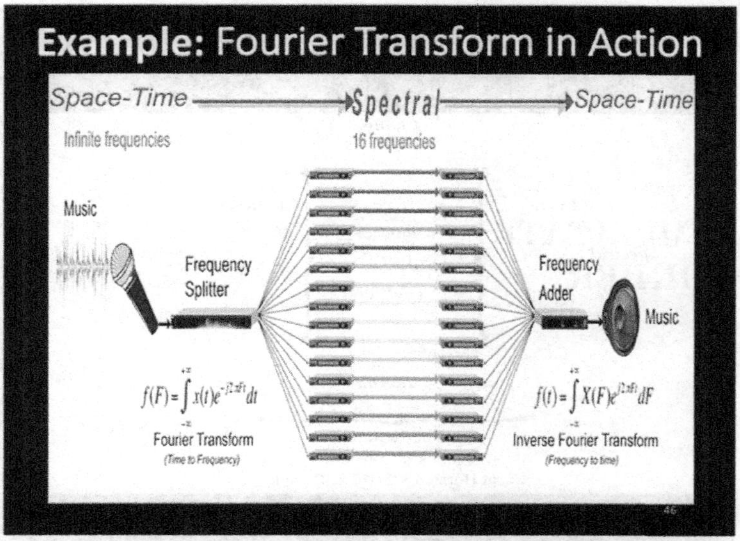

Here we see an engineering application of Fourier Transform processing.

Incoming music is detected by a microphone at the left.

At regular intervals, the acoustic wave
is sliced by a frequency splitter into 16 unique frequencies.

The amplitude of each frequency is recorded and stored.

Thus incoming **Space-Time music** has been converted to **Spectral** data.

The frequency data is then later unpacked, mixed,
and emerges from the speaker at the right, once more in **space-time**.

THE HOLOFLUX HYPOTHESIS posits that *the energy of consciousness* operates in a somewhat similar fashion, **continually transforming between** two domains.

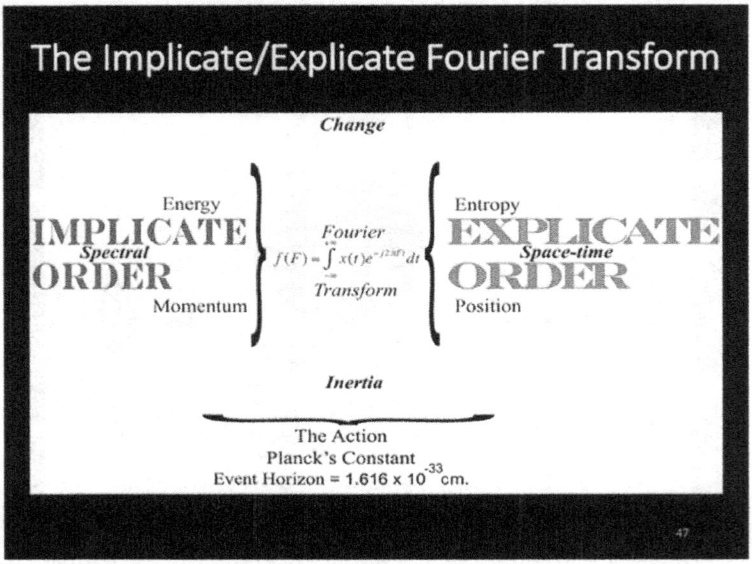

Here we see Bohm's IMPLICATE and EXPLICATE ORDER within the context of Dirac's diagram.

The Implicate Order lies within the SPECTRAL ENERGY REGION.

The Explicate Order lies within Space-Time to the right.

Again, at the **Center** we see the Fourier Transform.

But to fully understand abstract relationships such as these, it is more useful to visualize the relationships geometrically.

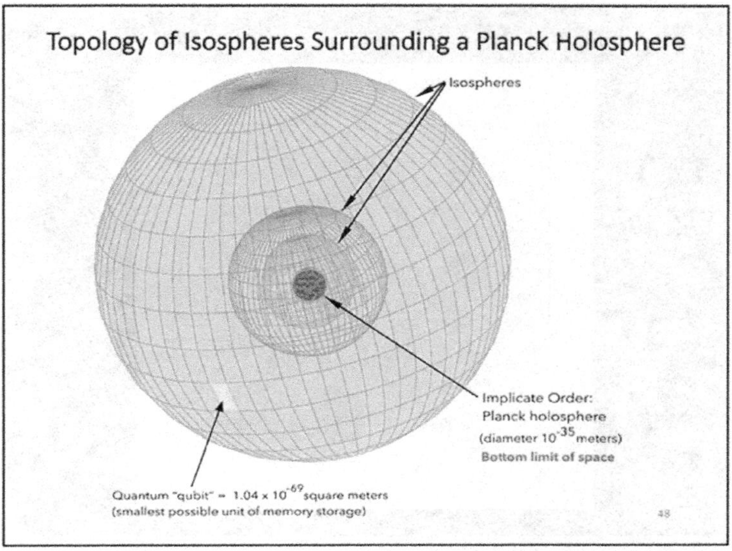

In this diagram we see a quantum black hole,
or Planck holosphere,
separating space from an interior Implicate Order.

Surrounding the central Planck holosphere are thin shells called isospheres, each providing enormous memory storage capacity.

For simplicity, only three isospheres are shown here,
but in actuality we should consider many,
one for each unique quantum frequency
in the virtually infinite
electromagnetic radiation spectrum.

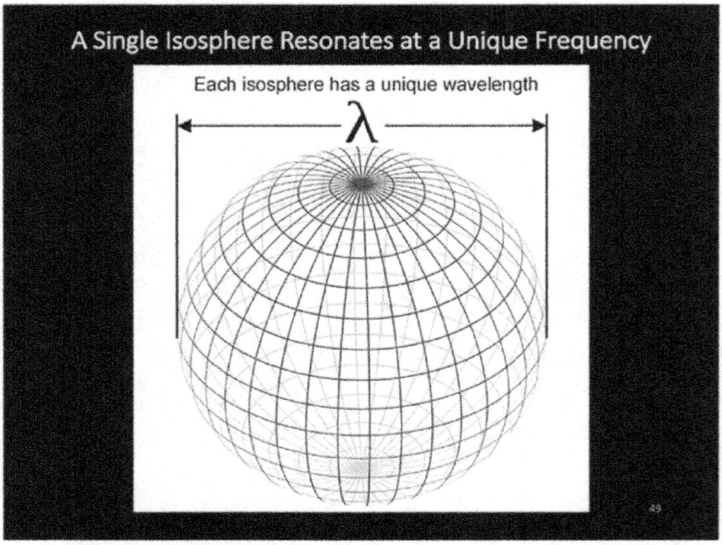

From RADIO THEORY, we know that each isosphere must resonate at the unique frequency
constrained by its specific diameter, its **wavelength**.

This unique frequency in the space-time **EXPLICATE ORDER** resonates with
the **IDENTICAL** frequency **within** the **IMPLICATE ORDER.**

This is how CONSCIOUSNESS within the nondual implicate order communicates with electromagnetic energy in our familiar space-time,
THROUGH FREQUENCY RESONANCE via the mathematics of the FOURIER TRANSFORM.

Notice the grid. Each rectangular grid represents ONE bit of information stored on the isosphere. Let me explain.

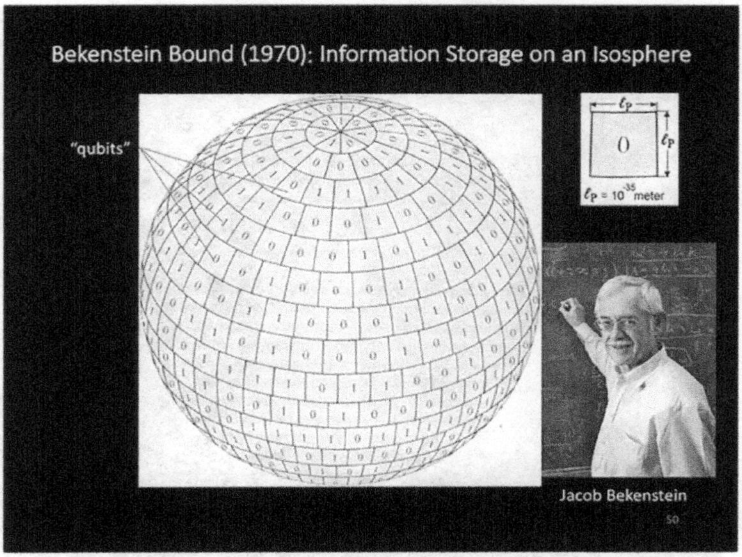

As a graduate student under John Archibald Wheeler, who coined the term "black hole," Jacob Bekenstein pioneered the mathematics of **black hole thermodynamics.**

Bekenstein defined the **qubit** as the smallest possible storage area in space, having an area of **one square Planck length**.

The upper limit to the information that can be contained upon the surface of a specific, finite volume of space has come to be known as the "Bekenstein bound."

And just why is this storage so important?

Because, along with
This recent Chinese teleportation experiment,
It implies that the universe may already have
a natural **faster-than-light**
Quantum Communications Network already in place.

In the experiment, photons were beamed
from a station in Tibet
to a Chinese satellite orbiting 300 miles above Earth.

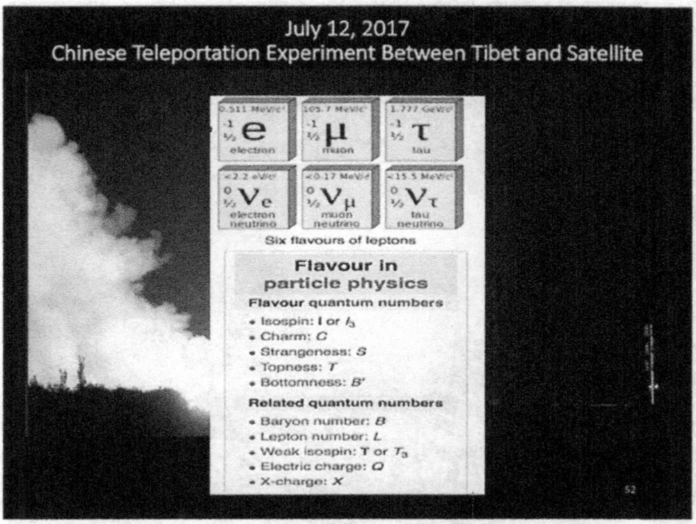

ALL of these quantum particle properties,
including FLAVOR values of Charm, Strangeness,
Topness, and Bottomness,
were instantaneously transferred
from a particle on the ground
to a particle 300 miles away
faster than the speed-of-light.
IN EFFECT
– PROVING that quantum TELEPORTATION may indeed by possible through the nonduality of the quantum black hole plenum below the Planck length,
via Bohm's IMPLICATE ORDER.

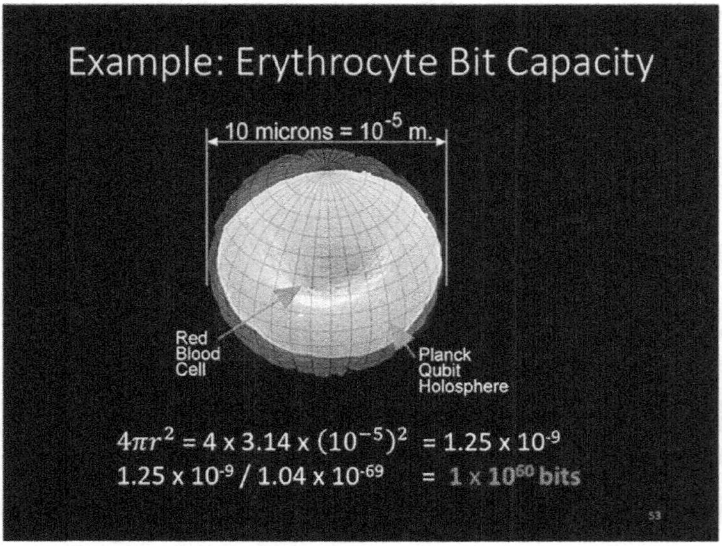

As an example OF HOLONOMIC STORAGE CAPACITY, particularly appealing to those who may consider blood as a possible conveyor of consciousness, let us examine the possible storage capacity of an Erythrocyte, or red blood cell, that mysterious cell
that has **no nucleus and does not reproduce itself,**
is filled with atoms of iron forming a ring, and
yet it is found ubiquitous throughout our body.

If we consider the diameter of a typical erethrocyte, about 10 microns,
the number of bits of information that can be stored on an isosphere bounding such a cell, is 10^{60} bits.

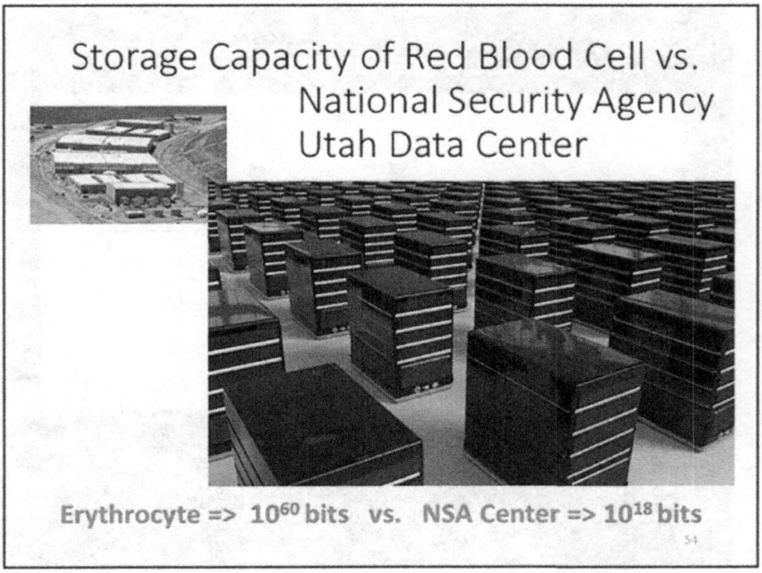

Note that by comparison,
the United States National Security Agency's **Utah Data Center** has a reported capacity of a mere 10^{18} bits of information.

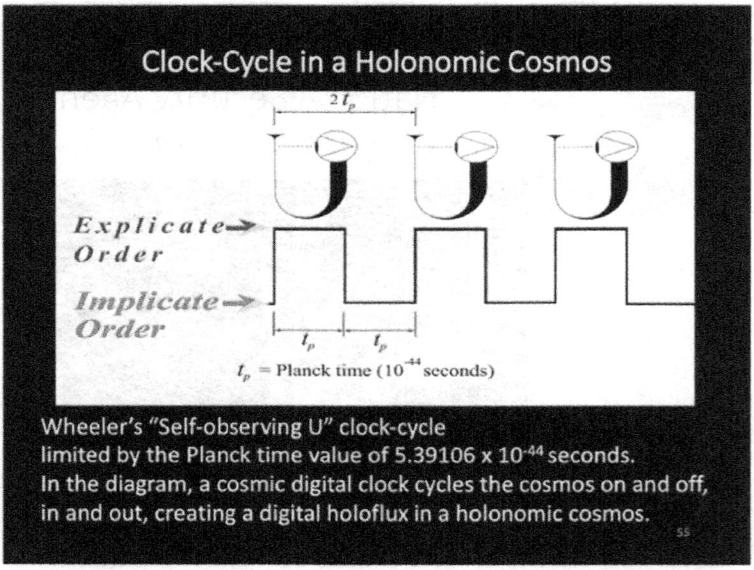

One final piece of the Pribram-Bohm topology concerns the inclusion of the Planck Time Constant into the equation.
How does the Planck TIME CONSTANT
influence the proposed holoflux process?

Using John Wheeler's famous "Self-Observing U," an image he often used as short hand to indicate the cosmology of the universe viewing its own creation,
this slide expresses the fundamental clock-cycle of the cosmos to be the Planck time of 10^{-44} seconds.

In the diagram, a cosmic digital clock cycles the cosmos
in and out, at a constant rate of the Planck time constant.

This is the digital processing of a holonomic cosmos
At a digital clock rate of
Ten to the forty-four pulses per second.

The implications here are enormous!
We can only begin to imagine the cosmos
as we know it
operating at such a clock cycle,

Imagine the entire plenum of galactic matter
collapsing
in and out between the implicate order
and
the explicate order
at a rate of 10^{44} cycles per second.

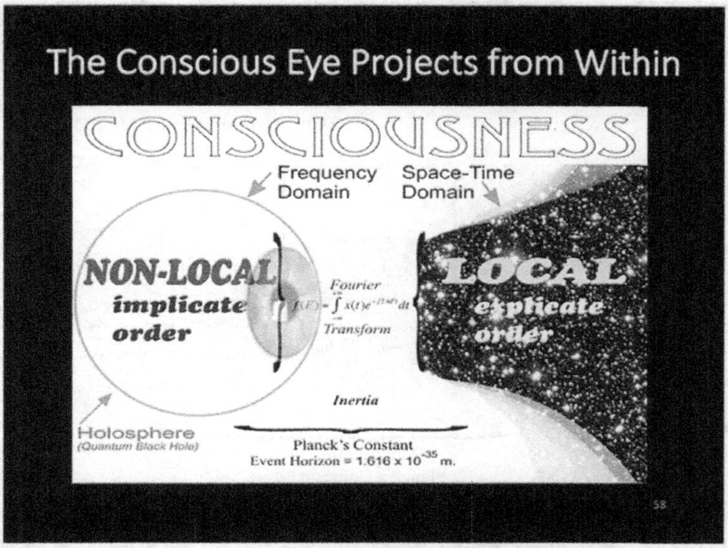

ANOTHER WAY of understanding this theory can be seen here,
AND HAS MAJOR IMPLICATIONS FOR CONSCIOUSNESS,

We have **here** over-laid the image of an **iris**
AT THE SURFACE OF THE
EVENT HORIZON OF AN IMPLICATE ORDER HOLOSPHERE,
TO THE LEFT IN THE DIAGRAM.

WE SEE **consciousness,**
from **within** the nonlocal implicate order
peering **out into** our local space-time **explicate order** of visible stars and galaxies.

BOHMS SAYS THAT space-time UNFOLDS FROM the IMPLICATE ORDER, that forms are PROJECTED **out into**
SPACE AND TIME
from WITHIN the **IMPLICATE ORDER.**

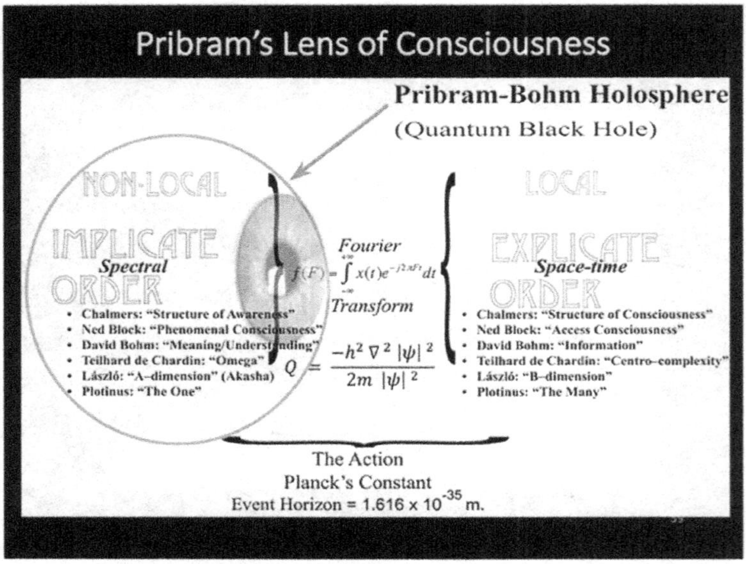

Bohm's holoflux plasma mathematics
fully supports such a topology.

You can see Bohm's equation for what he called the QUANTUM POTENTIAL, ***Q***, in the center of the diagram, right below the Fourier Transform and above The Action.

Karl Pribram was in full agreement with Bohm, and summarizes his own thinking here,
at the age of 94,
and I quote:

"These two domains characterize the input to and output **from a lens**
that performs a **Fourier transform**."

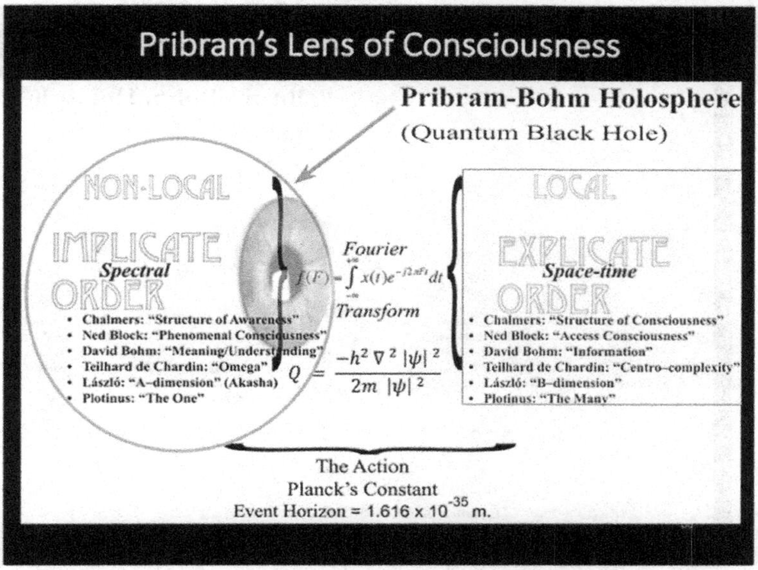

On one side of the transform lies the space–time order we ordinarily perceive.

Bohm's **Explicate Order**.

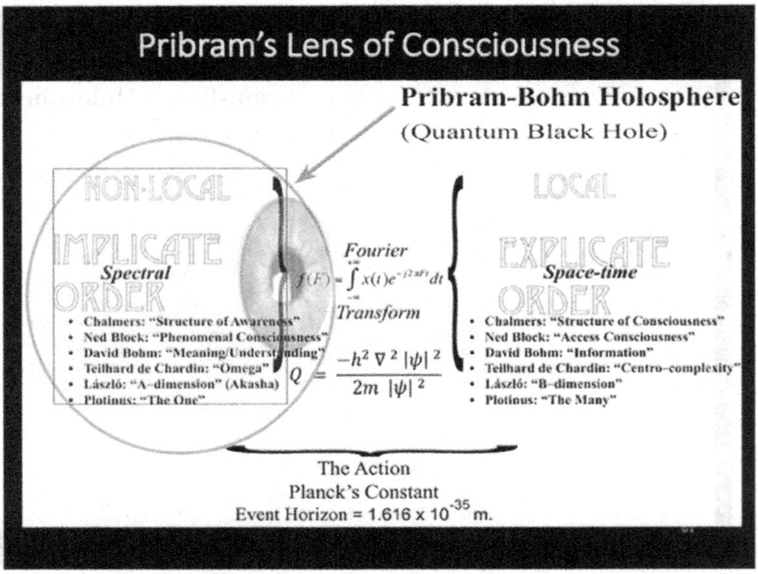

And on the other side lies a distributed enfolded holographic-like order referred to as the frequency or spectral domain," Bohm's **Implicate** .

What might be the implications of the Pribram-Bohm theory, particularly for **consciousness studies, paranormal research, and non-dual experiences**?

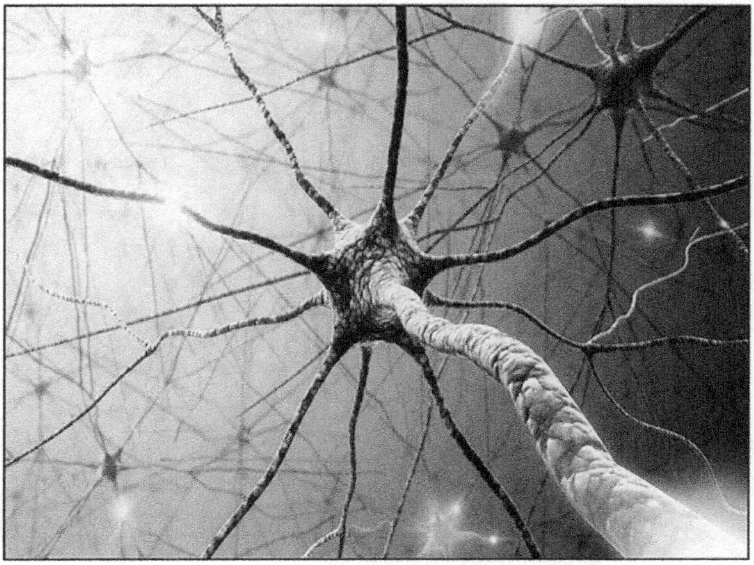

The implications are that consciousness, the experience itself,
is **not** an EPIPHENOMENON of **nerve impulse powered** neurons,

Instead, if we follow the lead of Pribram and Bohm,
consciousness should be regarded as **primary**,
as radiant energy operating lens-like
to focus and tune between the frequency domain and space-time.

AND WHAT might this imply for a human physiology of consciousness?

WHERE DO WE LOOK THEN, TO IDENTIFY structures within the human body
capable of channeling high frequency conscious radiation?

BUT **FIRST** WE NEED TO TAKE A SHORT DIVERSION HERE INTO THE WORLD
OF RADIO ENGINEERING ANTENNA THEORY.

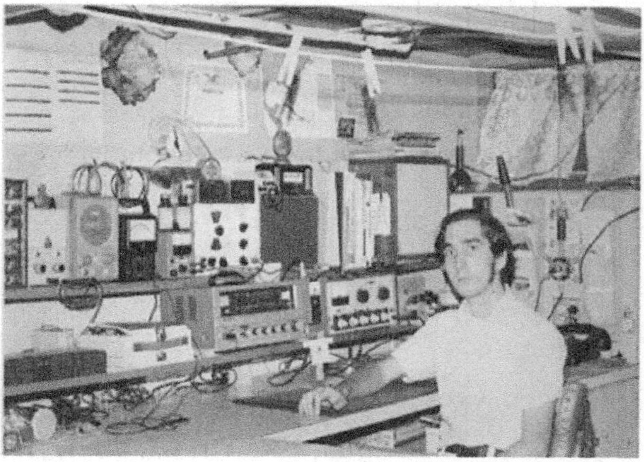

LET US TAKE A MINUTE

TO REVIEW THE BASIS OF RADIO COMMUNICATION

USING INVISIBLE ELECTROMAGNETIC WAVES.

THE FUNDAMENTAL LINK INVOLVES
WHAT IS CALLED THE ANTENNA.

Classical Antenna Theory

Maximum voltage differential
is between crest and trough,
or one-half of the wavelength.

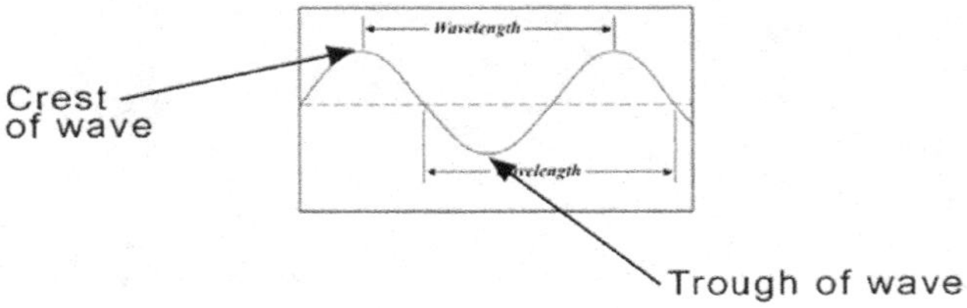

ANTENNA THEORY TELLS US THAT FOR MAXIMUM SENSITIVITY,
BOTH FOR RECEIVING AS WELL AS FOR BROADCASTING,

THE PHYSICAL ANTENNA WAVELENGTH MUST BE **ONE-HALF**
OF THE SIZE OF **THE TARGET FREQUENCY'S WAVELENGTH.**

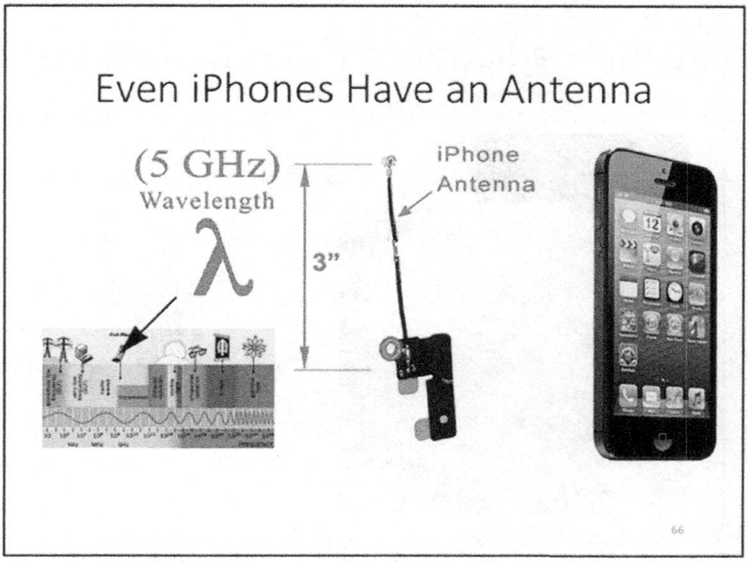

HERE IS AN EXAMPLE

An iPhone signal has a **wavelength**
of approximately 6 inches.

For this reason, the internal antenna of an iPhone
is **one half wavelength, or 3 inches.**

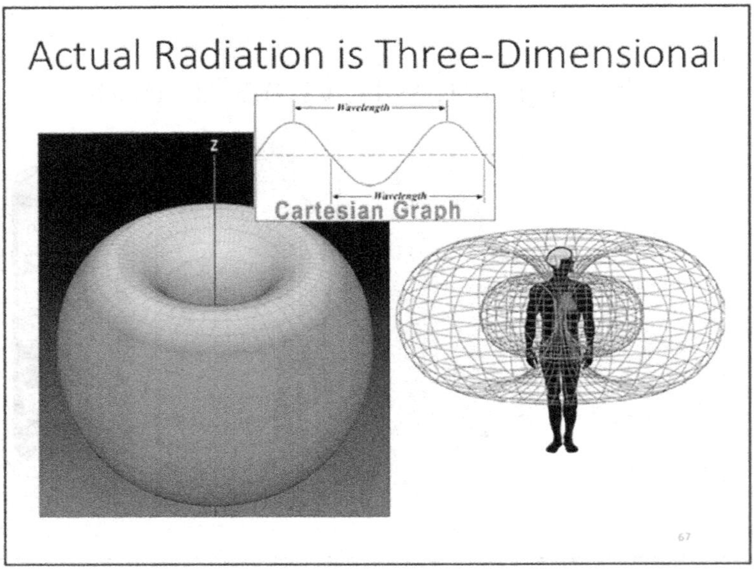

BUT THE CARTESIAN LINE **GRAPH OF A SINE WAVE IS SIMPLISTIC AND MISLEADING.**

ACTUAL RADIATION IS "THREE-DIMENSIONAL,"
AS IT
MOVES OUT FROM THE ANTENNA
OR FROM THE HUMAN HEART
AT THE SPEED OF LIGHT.

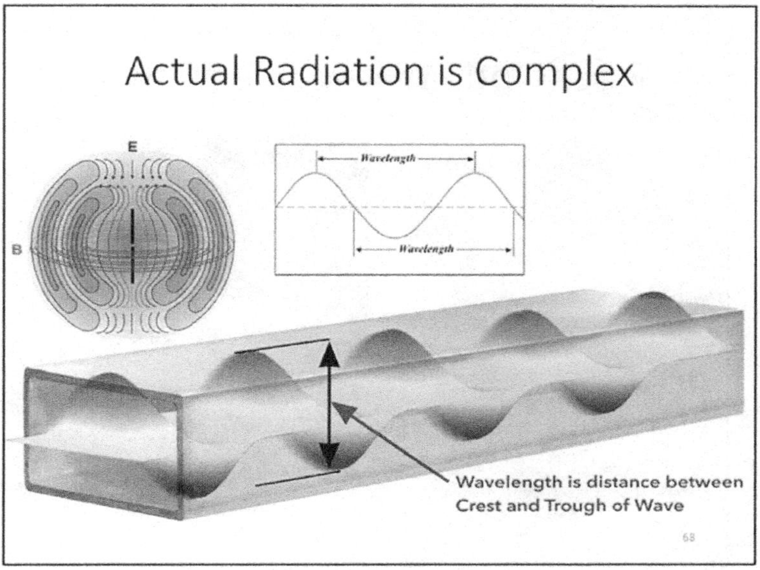

HERE ARE **THREE DIFFERENT WAYS**
TO VISUALIZE RADIATION.

AT THE CENTER WE SEE A TYPICAL SINE WAVE GRAPH.

TO THE LEFT IS A MORE REALISTIC, CROSS-SECTIONAL
IMAGE OF THE SAME SINE WAVE.

BUT ALONG THE BOTTOM WE SEE A THREE-
DIMENSIONAL VIEW OF
A SINE WAVE CONSTRAINED WITHIN A **WAVEGUIDE.**

THE DIAMETER OF A WAVEGUIDE ACTS
TO GUIDE THE WAVE FORWARD
AND TO **PROTECT** IT FROM OUTSIDE **INTERFERENCE.**

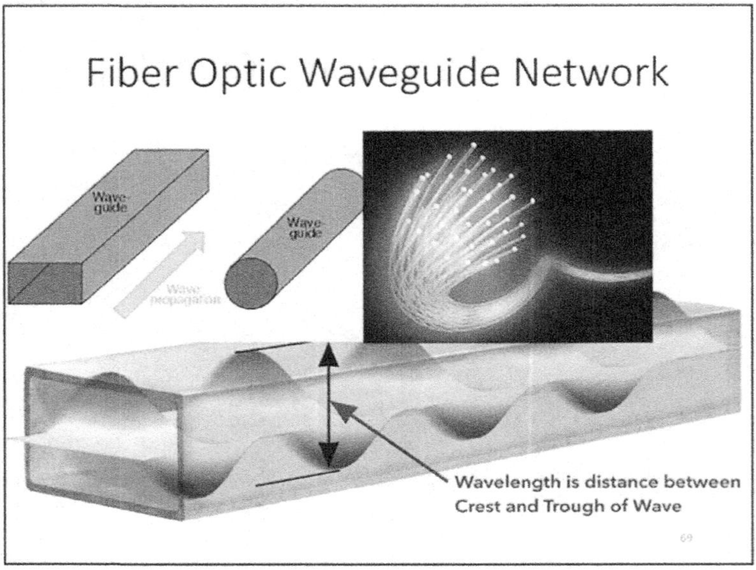

THE **INTERNET** IS A NETWORK OF FIBER OPTIC WAVEGUIDES.

THE SUM TOTAL OF INTERNET
FIBER OPTIC WAVEGUIDES STRETCHES FOR OVER **550,000** MILES AROUND THE GLOBE.

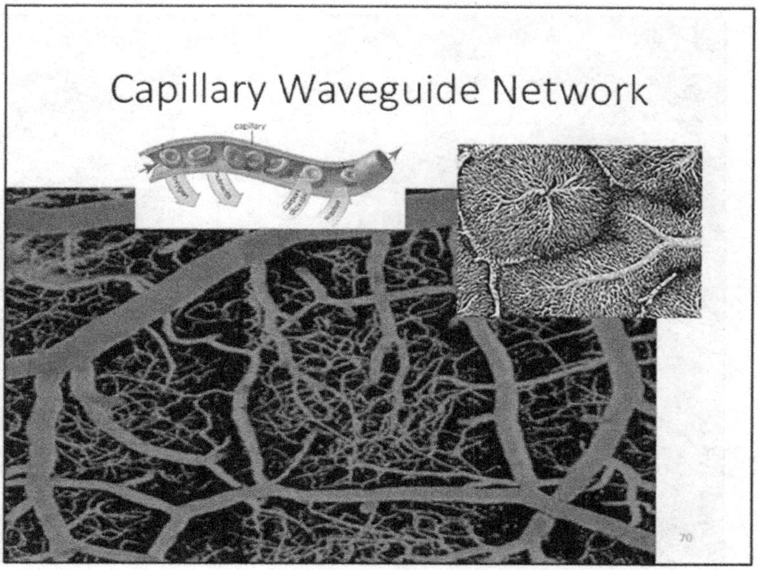

By contrast, the human capillary system
forms a network that
stretching for more than 59,000 miles
throughout the human body (and brain).

It offers a perfect network waveguide
for infrared signals,
and is ideal for 10 microns radiation,
the average diameter of capillaries.

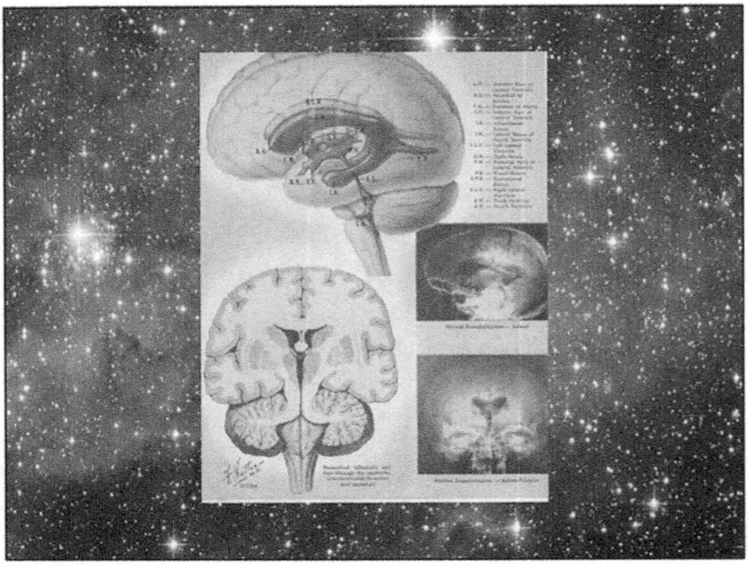

AND this image from a medical textbook
shows the hollow **ventricular cavities**,
filled with clear cerebrospinal fluid.

Medical textbooks say that the only function of these cavities is to help cushion the brain from trauma impact due to concussion.

BUT DOES IT SEEM PLAUSIBLE that such beautifully streamlined shapes, appearing to be
so similar to microwave horn antenna waveguides
should have evolved in the center of the cranium
merely to cushion us from **blows** to the head?

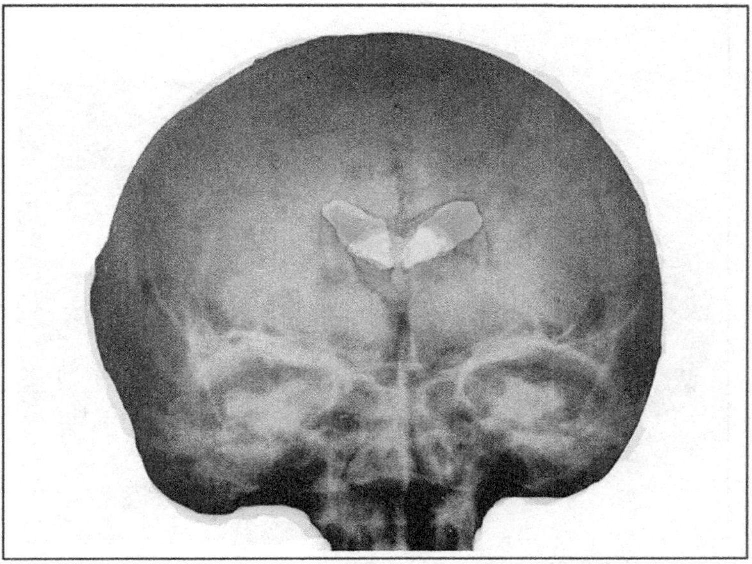

And this x-ray of a human skull
REVEALS THE **WINGED-SHAPE**
OF THE ventricular cavities
AS SEEN FROM THE FOREHEAD.

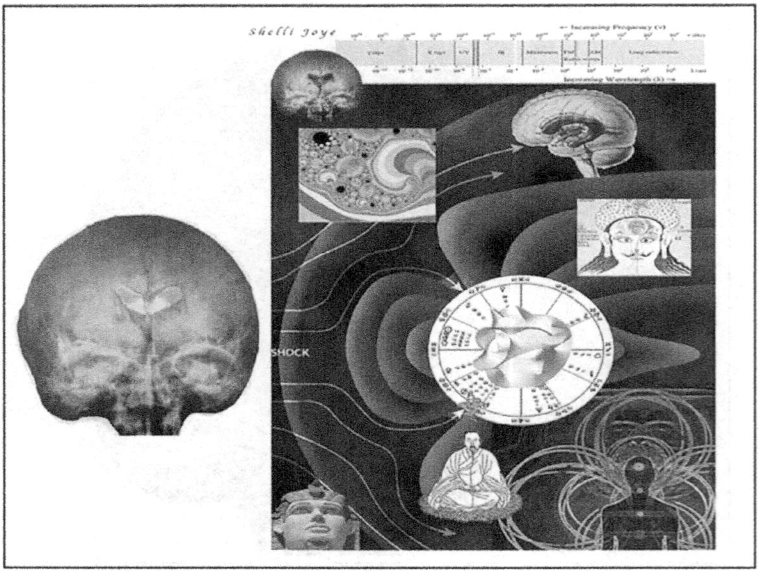

IS IT ONLY A COINCIDENCE
THAT THESE CAVITIES APPEAR
in the **PRECISE** location depicted by
illustrations of **"the third eye,"**
IN EGYPT, CHINA AND INDIA???

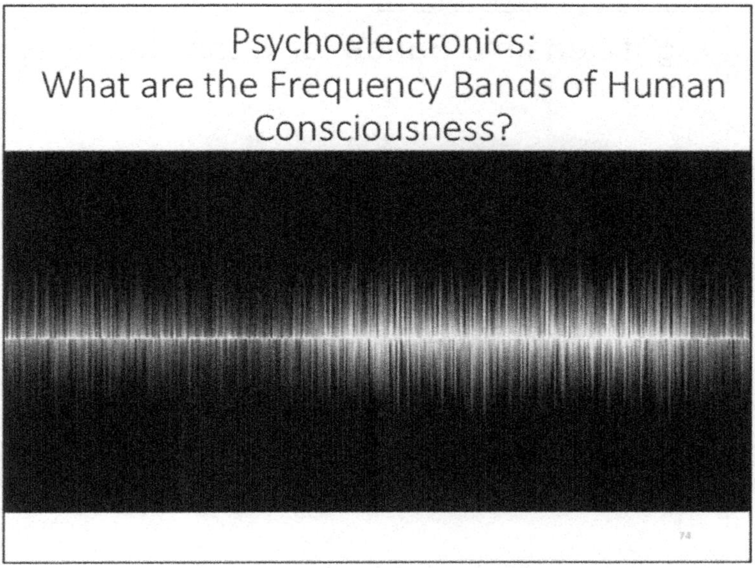

NOW IF THE THEORY IS THAT THOUGHT INVOLVES **TUNING THE ELECTROMAGNETIC FREQUENCIES OF THE BRAIN WITHIN THE HUMAN CAPILLARY WAVEGUIDE SYSTEM,**
THE NEXT QUESTION MUST BE,

"WHERE MIGHT BE THE VARIOUS **FREQUENCY BANDS** TO WHICH WE MIGHT TUNE OUR CONSCIOUSNESS?"

WHAT ARE THE VARIOUS RANGES AND BANDWIDTHS OF HUMAN CONSCIOUSNESS?

THIS IS IMPORTANT BOTH FOR
FUTURE RESEARCH IN THE PHYSICS OF CONSCIOUSNESS
AS WELL AS FOR THE PIONEERING
CONTEMPLATIVE EXPLORATION BY PSYCHONAUTS.

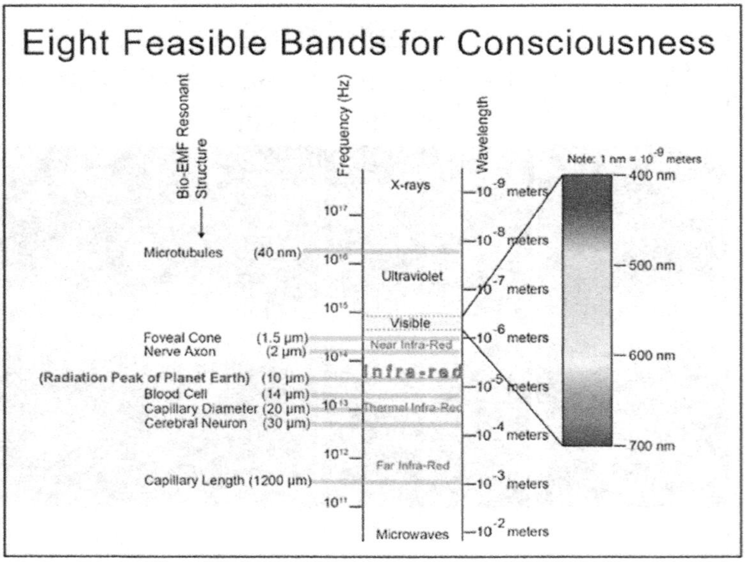

THIS DIAGRAM INDICATES POSSIBLE CHANNELS FOR CONSCIOUSNESS WITHIN THE HUMAN BODY.

Here we compare the size of various structures
To see where electromagnetic wavelengths will
resonate according to antenna theory.

Note that the band of **visible light wavelengths** fall somewhere in the central range.

At the highest frequencies, we see microtubules providing waveguides in ultraviolet frequency bands, just slightly above visible light.

Below the visible light band, we see waveguide possibilities in blood capillaries channels in the infrared frequency band.

Here we see the Japanese researchers
Mari Jibu, and Kunio Yasue (with Stuart Hameroff).
Jibu and Yasue extended the findings of the physicist
Hiroomi Umezawa to develop
a theory called **Quantum Brain Dynamics** (QBD).

In the 1990s they worked closely with Karl Pribram, publishing their physiological mathematical theory in an Appendix to one of Karl's books to provide a sound mathematical basis for Karl's holonomic brain theory.*

In the upper left is a model of the magnetic field, generated from **one dipole** of a water molecule.
In the upper left, we see a molecular lattice which forms a plenum of aligned water molecule dipoles which offer the basis of what is termed **the cortical field**.

---

* Jibu and Yasue, "A Theory of Nonlocal Cortical Processing in the Brain," in Pribram's *Brain and Perception: Holonomy and Structure in Figural Processing.*

The Electromagnetic Cortical Field

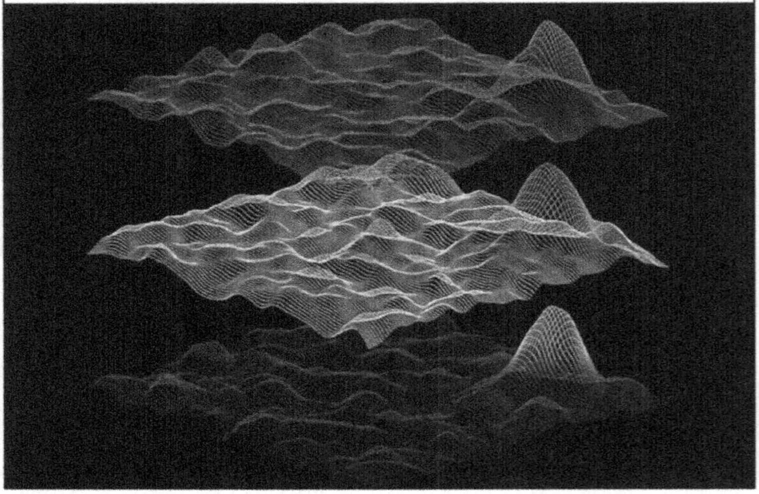

According to Jibu and Yasue,
the combined electric dipoles
of **all** the water molecules in the
human blood stream
constitute a **single quantum field,**
which resonates within a narrow
band of frequencies in the far infrared.

In this model of consciousness, the human blood system acts as a **single giant water molecule resonating as one system of ionic plasma.**

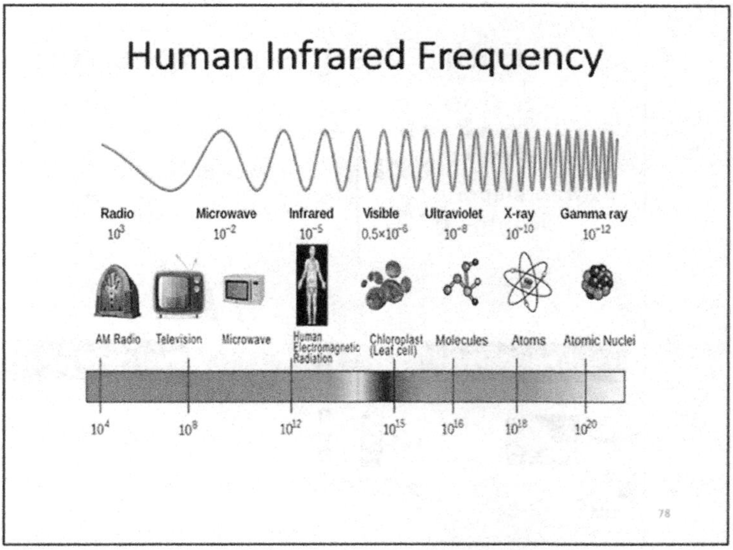

In this figure we can view a small band of the electromagnetic spectrum,
from radio waves at the left to gamma rays on the right.

Note that **as the wavelength become physically smaller,**
(as we see as the axis moving toward the right)
the frequency rate increases.

The physical size of a wavelength is inversely proportional to the frequency.

Higher frequencies, capable of encoding more information, are found at smaller dimensions.

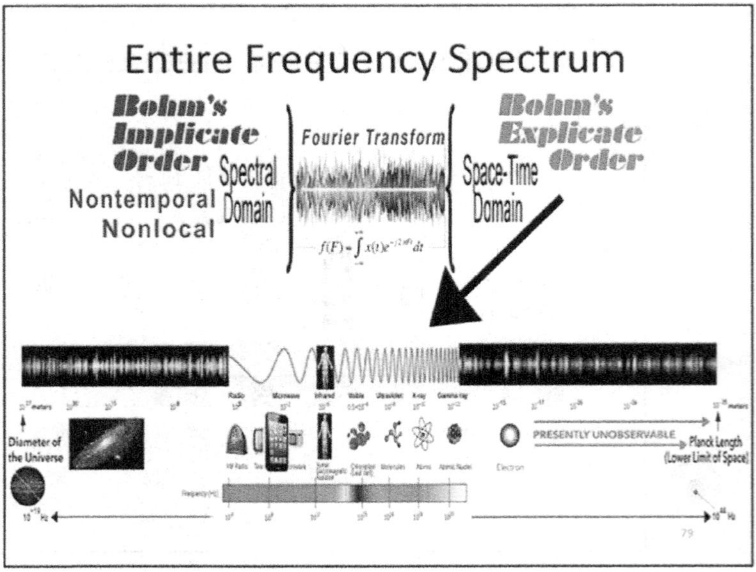

Here we view the same range of frequencies
in the explicate order's electromagnetic waves
inserted into the center of the **entire FREQUENCY SPECTRUM RANGE OF THE UNIVERSE**,
which can be seen here ranging from
the diameter of the entire universe at the far left
end of the scale (at value of $\mathbf{10^{27}}$ meters),

down to the tiny Planck length limit ($\mathbf{10^{-35}}$ meters)
located at the right end of the scale.

In terms of frequency, the universe ranges from
$\mathbf{10^{-19}}$ Hz to $\mathbf{10^{44}}$ Hz.

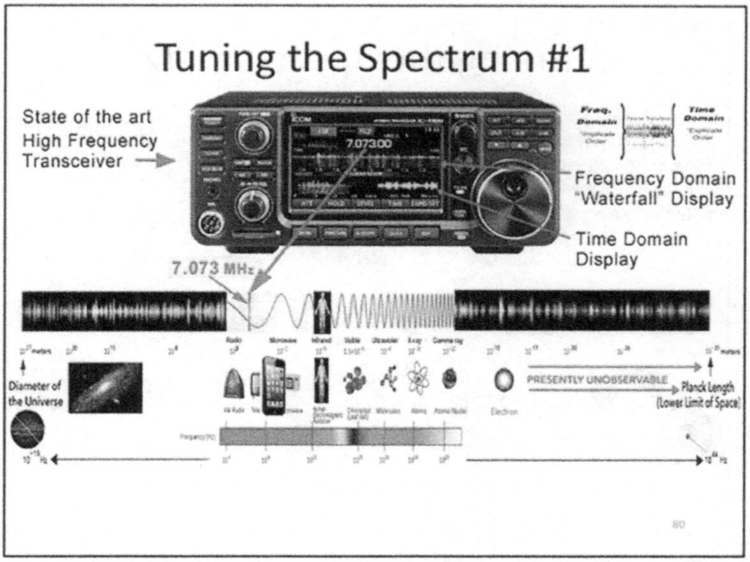

This is a typical modern communications **radio transceiver**, used both **to receive** and **to transmit** electromagnetic energy, but only within a very small range of the vast electromagnetic spectrum.

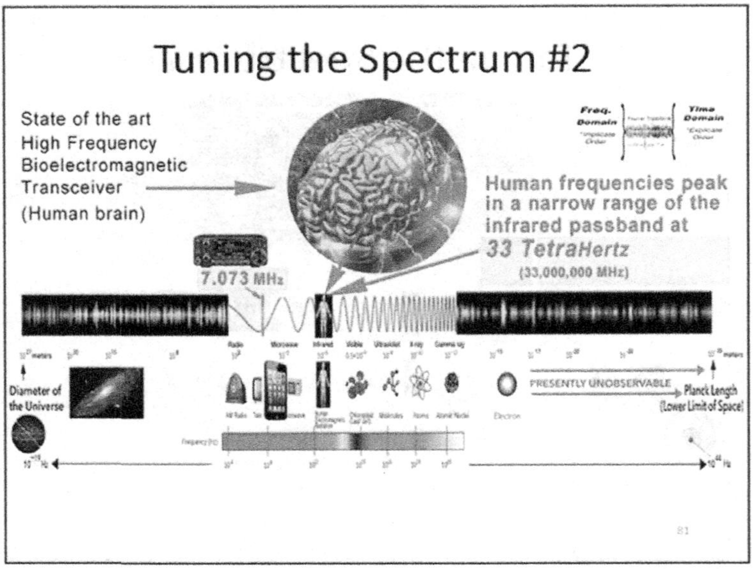

By contrast, Pribram's holonomic brain,
viewed as a transceiver,
may be tuned to the infrared band,
just below the visible light frequencies.

Human infrared peaks at 33 Terahertz,
much higher than radio frequencies.
offering rich computational bandwidth
for thought, sound processing,
and image visualization.

But what **evidence** do we have
that the human brain might operate
in the infrared band region?

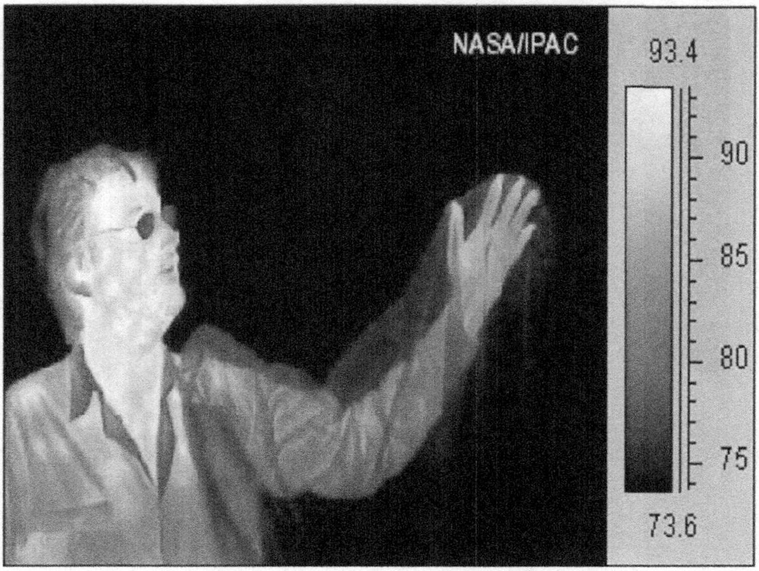

FIRST OF ALL,
HUMANS **GLOW** WITH **INFRARED FREQUENCIES.**

THIS IS A PHOTOGRAPH OF ELECTROMAGNETIC ENERGY RADIATING FROM THE HUMAN BODY, AS SEEN THROUGH "NIGHT VISION GOGGLE" TUNED TO THE INFRARED BAND

WITH AN OUTPUT OF 860 WATTS THE HUMAN BODY **GLOWS** WITH ELECTROMAGNETIC ENERGY, THOUGH THE PHENOMENON IS COMMONLY CALLED "HEAT."

THE **SCALE** ON THE RIGHT **PLOTS TEMPERATURE** VS **LUMINOSITY** ACCORDING TO **WIEN'S LAW.**

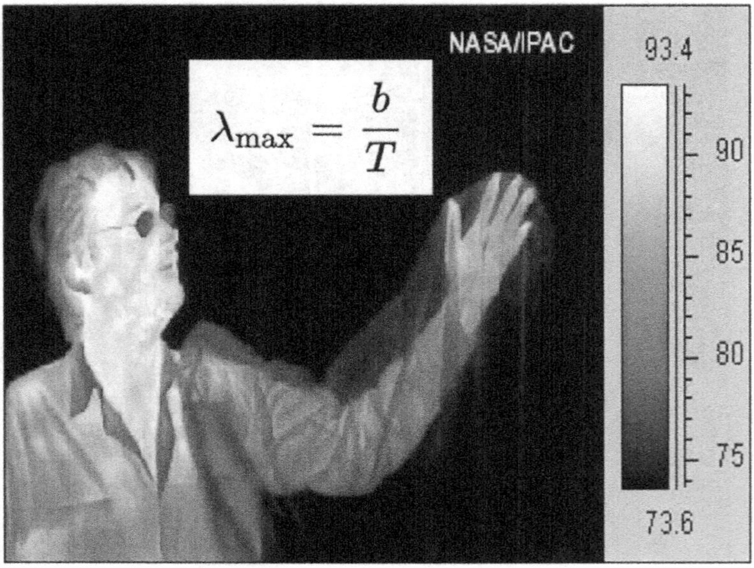

Wien's Law is an equation that states that
the MAXIMUM WAVELENGTH
Of ELECTROMAGNETIC RADIATION varies with the
INVERSE OF THE TEMPERATURE.

WITH THIS EQUATION WE CAN calculate
the **PEAK wavelength**
radiating at **ANY given** specific temperature **T**

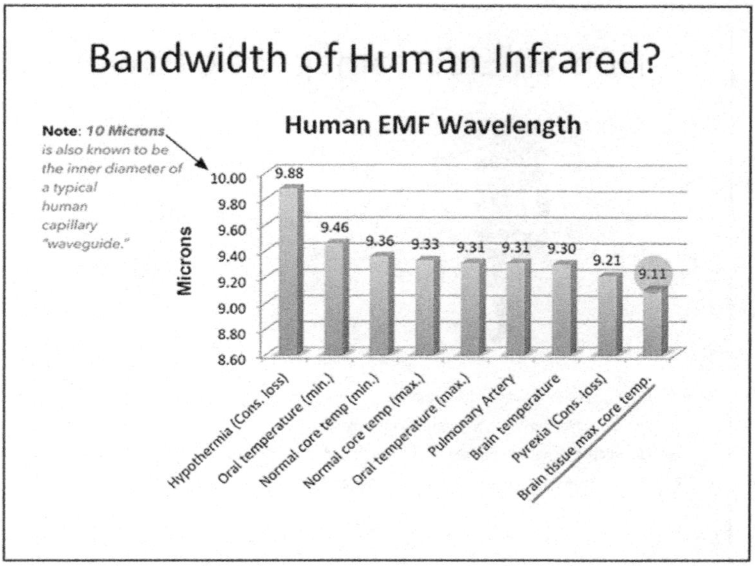

SO WHAT MIGHT BE THE
BANDWIDTH OF HUMAN INFRARED RADIATION?

Applying Wien's Law
at the location of various internal body temperatures,
this chart reveals
the bandwidth of human infrared
to lie close to 10 microns.

In terms of frequency this is
in the 33 Terahertz frequency band, ideal for
high speed information, image, audio, and thought
processing.

Human Electromagnetic Power

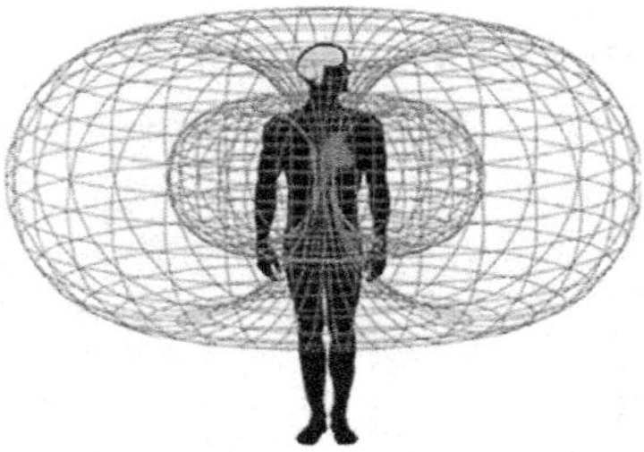

	Output Power
Energy radiated by most powerful radio on the planet	1,500,000 watts
Energy radiated by nuclear reactor (Three Mile Island @ maximum)	873,000 000 watts
Energy radiated by human heartbeats (combined @ 7 billion)	9,100,000,000 watts

In terms of power,
each human heartbeat radiates
a total of **1.3 watts** of infrared electromagnetic energy.

An **iPhone** transmits **1.6 watts** of electromagnetic energy.

And the most powerful radio on the planet broadcasts
a maximum of **1.5 million watts**.

But by contrast,
the combined output of 7 billion human heartbeats is
an astounding
9.1 Gigawatts of electromagnetic infrared power
broadcasting into the noosphere, continuously.

Resonance in the Thermosphere?

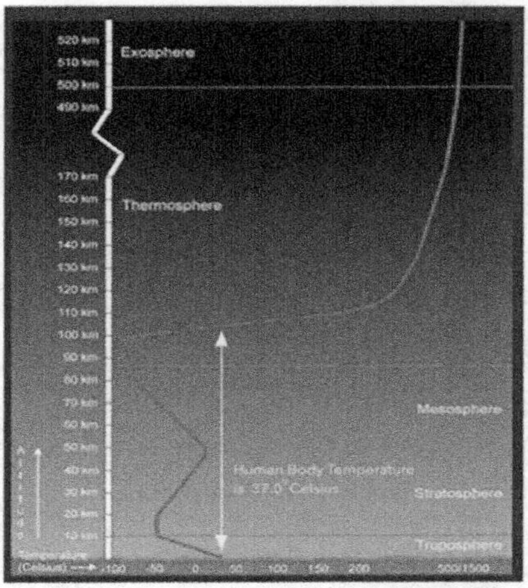

In this graph
we see temperature at various distances from the ground.

Notice how temperature
at first **decreases** as we rise in altitude,

But for reasons not well understood,
the temperature increases sharply above 90 km

At 100 kilometers above sea level the atmospheric
temperature passes through a zone that mirrors
human body temperature at 37 degrees Celsius.

Resonance in the Noosphere?

Could cavity resonance in the thermosphere
be the locus of what Teilhard de Chardin
and Vladimir Vernadsky
termed the **Noosphere***?

* In the theory of Vernadsky, the **Noosphere** is the third in a succession of phases of development of the Earth, after the Geosphere (inanimate matter) and the Biosphere (biological life). For Teilhard de Chardin, the Noosphere is the sphere of thought encircling the earth that has emerged through evolution as a consequence of this growth in complexity/consciousness.

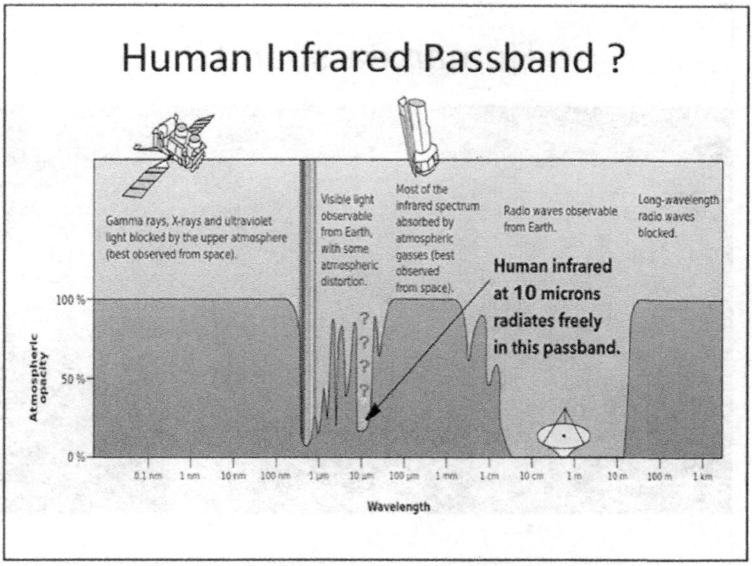

The atmosphere is opaque and **blocks** radiation at most wavelengths.

However this plot reveals major **passbands** in atmospheric opacity.

As we all can attest, nature has evolved human organs of perception for **the passband** we call **visible light.**

But what about the distinct passband shown here at around 10 micron, the human infrared range?

Human Passband ?

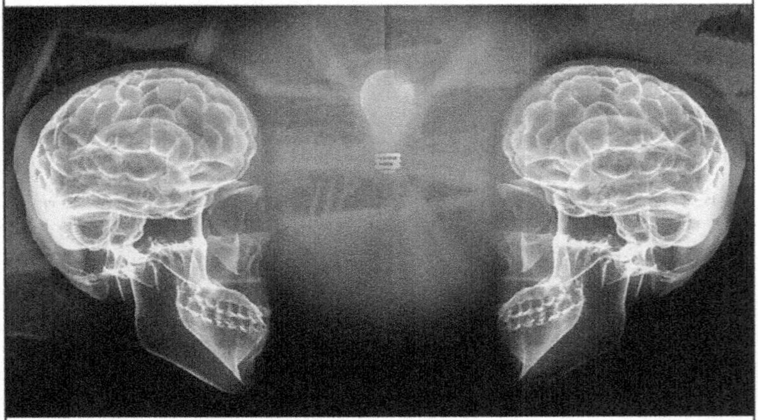

What **organs of perception** might nature have evolved to take advantage of the passband at 10 microns?

Could this be the doorway to regions explored by mystics, yogis, shamans, and saints, and psychonauts throughout recorded history?

Let us take a moment to consider what these ideas imply.

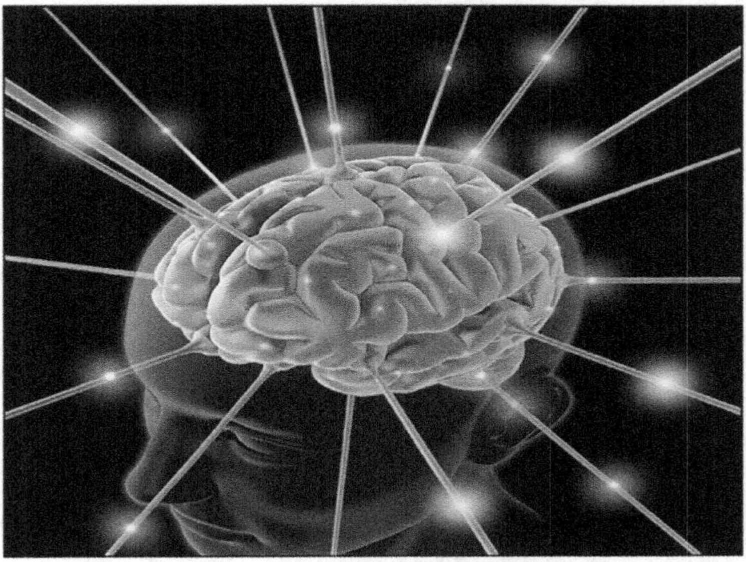

The neocortex is in a way like a personal computer

We can learn
to turn down its activities of verbalization and memory.

We do this in order to allows the deeper consciousness
to become aware of supersensible and normally
inaccessible regions of awareness.

By halting the normal flow of thoughts and memory,
we can suspend our normally active neuronal
biocomputer, putting it into sleep mode.

Only then can we begin to perceive what is beyond
the normal waking boundaries of awareness.

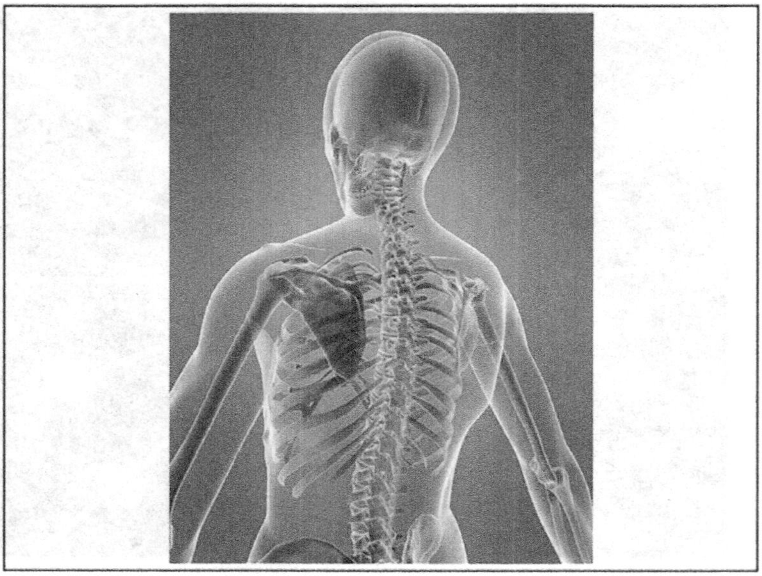

Among those who practice silent contemplation
to explore consciousness, among the
monks, shamans, mystics, and psychonauts,
one widely reported experience is a feeling of
electromagnetic energy that is often
perceived to be rushing upwards,
moving up along the central channel of the spine.

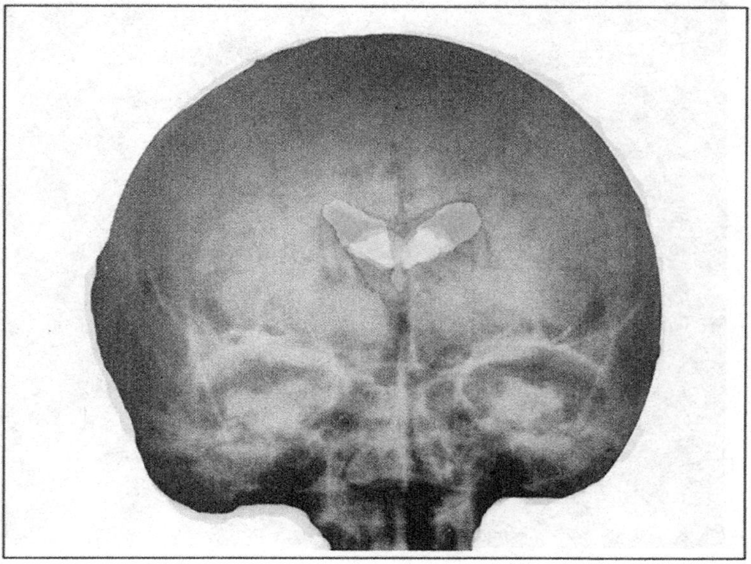

Recalling our x-ray of the ventricular cavities
we see that they are in the **exact** location depicted
by occult illustrations of "the third eye'
experienced by contemplatives,
and called in Sanskrit the
***Brahmarandra* or "the cave of Brahma."**

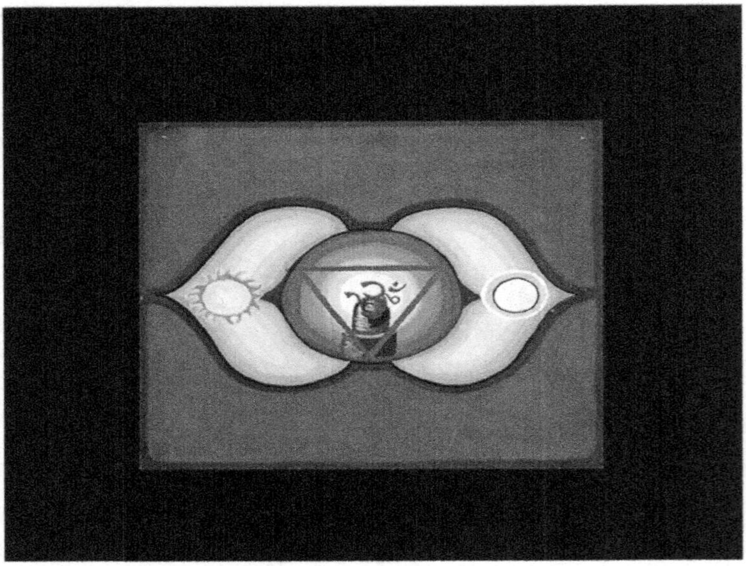

Here we see a
12th century Kashmiri diagram of **the third eye**,
appearing to be the same shape and in the same location as
the **ventricular cavities**
which we saw through the forehead x-ray.

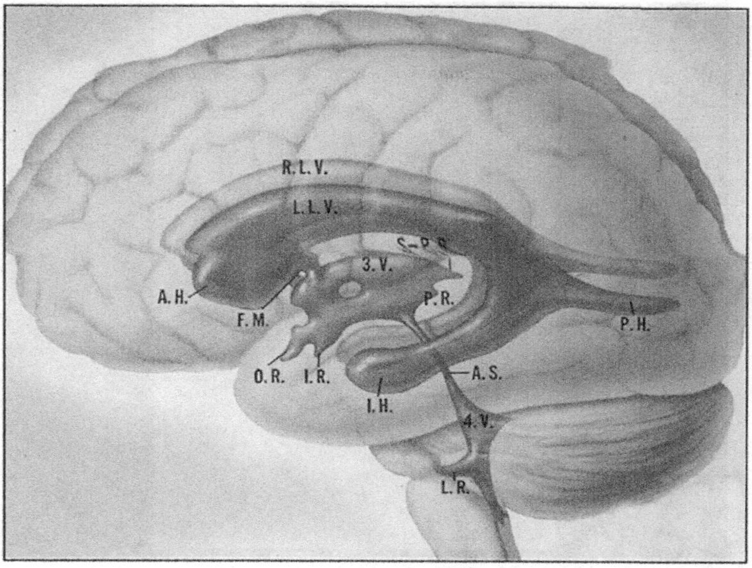

During normal waking activity,
when the cerebral hemispheres are electrically active,
the deeper inner areas are **shielded**
due to
the **Faraday cage effect**.

In a Faraday cage
the moving electrical charges on the outer regions shield
the inner regions.

However when the cortex is silent, such as during sleep,
or during silent contemplation,
the ventricular cavities
are suddenly open to signals coming through
from the outside universe.

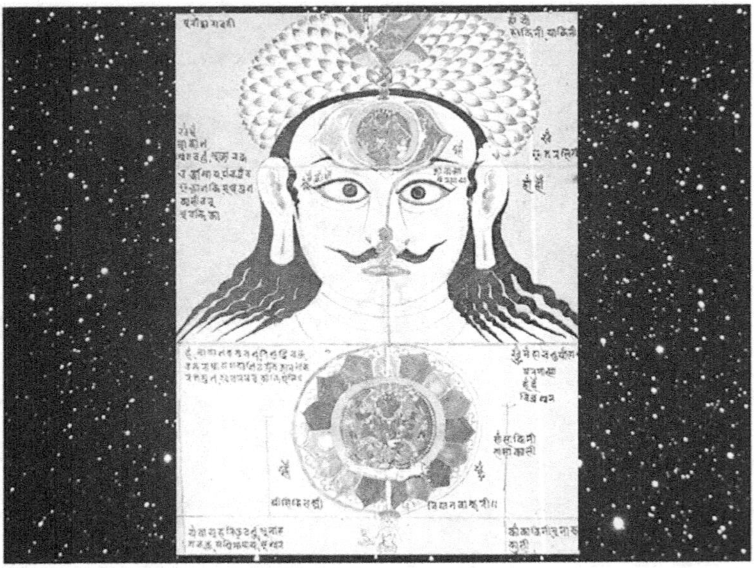

Another diagram from Rajasthan showing the third eye chakra at the location of the ventricular cavities.

Also depicted here is the heart chakra.

The heart has been and continues to be
An important center for practicing and developing focus of consciousness.

In fact,
there are **innumerable** locations within the human body that can be developed through contemplative practices such as yoga, Hesychasm, or psychonautic exercises.

This diagram shows many areas within the body
that can be cultivated through
Samyama, according to the instructions found in the classic text on contemplative practices, Patañjali's ***Yoga Sutra***.

Patañjali tells us that, after quieting the normal monkey activity of the brain/mind, one should focus upon
an internal physiological center such as the heart or the abdomen, or the fontanelle at the top of the head, and by sustaining the concentration for a period of time, new sensations will be evoked and the center will grow in sensitivity, eventually allowing supersensible capabilities (in Sanskrit called *siddhis*) to develop.

This image is taken from an ancient scroll book on Taoist Chinese meditation.

According to the Taoist texts,
when the mind is sufficiently stilled,
the **true self** is seen to rise out of the cranium through the fontanelle and travels up and out to commune with the living cosmos,
perhaps tuning in
to what
Teilhard de Chardin referred to as
"the evolving noosphere."

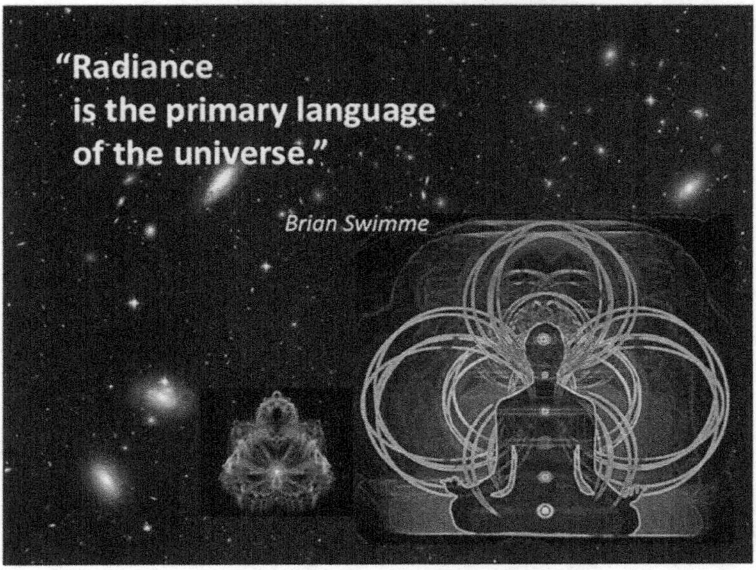

In closing, I would like to share
a relevant quote from my mentor
and fellow explorer,
the mathematical cosmologist
Brian Swimme, who has often said

"RADIANCE IS THE PRIMARY LANGUAGE OF THE UNIVERSE"

Endnotes

[1] Graphic by Darth Kule (2010). Public domain. Image retrieved from Wikimedia Commons. https://commons.wikimedia.org/wiki/File:Black_body.svg

[2] Bohm, *Wholeness and the Implicate Order*, 188.

[3] Bohm, 189.

[4] Bohm, 189.

[5] Graphic by Camille Flammarion (Paris, 1888). Reprinted under the terms of a Creative Commons Attribution ShareAlike 3.0 Unported license. Image retrieved from Wikimedia Commons.

[6] Graphic by J. Wheeler (Austin, 1990). Reprinted under the terms of a Creative Commons Attribution ShareAlike 3.0 Unported license. Image retrieved from Wikimedia Commons.

[7] Ibid., 142.

[8] Pribram, "What the Fuss Is All About," 29.

[9] Pribram, "Prolegomenon for a Holonomic Brain Theory."

[10] Pribram, "Brain and Mathematics," 232.

[11] Pribram, *The Form Within*, 495.

[12] Kuo, *Network Analysis and Synthesis*, 1.

[13] Pribram, *Brain and Perception*, 73.

[14] Stein and Shakarchi, *Fourier Analysis*, 134–36.

[15] Kuo, *Network Analysis and Synthesis*, 40.

[16] Wiener, *Cybernetics: Control and Communication in Animal and Machine*, 198.

[17] Ibid.

[18] Ibid., 202.

[19] Ibid. Italics added.

[20] Mandelbrot, "Fractals and the Rebirth of Iteration Theory," 151.

[21] Penrose, *The Emperor's New Mind: Concerning Computers, Minds and the Laws of Physics*, 124.

[22] Ibid., 127.

[23] Jung, "On the Nature of the Psyche," 215.

[24] Kuo, *Network Analysis and Synthesis*, 13.

[25] Ibid.

[26] Ibid.

[27] Ibid., 3.

[28] Bohm, *Quantum Theory*, 1.

Made in the USA
Monee, IL
04 November 2026

67384734R00115